The **Void**

The **Void**

FRANK CLOSE

OXFORD
UNIVERSITY PRESS

OXFORD
UNIVERSITY PRESS

Great Clarendon Street, Oxford OX2 6DP

Oxford University Press is a department of the University of Oxford.
It furthers the University's objective of excellence in research, scholarship,
and education by publishing worldwide in

Oxford New York

Auckland Cape Town Dar es Salaam Hong Kong Karachi
Kuala Lumpur Madrid Melbourne Mexico City Nairobi
New Delhi Shanghai Taipei Toronto

With offices in

Argentina Austria Brazil Chile Czech Republic France Greece
Guatemala Hungary Italy Japan Poland Portugal Singapore
South Korea Switzerland Thailand Turkey Ukraine Vietnam

Oxford is a registered trademark of Oxford University Press
in the UK and in certain other countries

Published in the United States
by Oxford University Press Inc., New York

British Library Cataloguing in Publication Data

Data available

Library of Congress Cataloging in Publication Data

Typeset by SPI Publisher Services, Pondicherry, India
Printed in Great Britain
on acid-free paper by
Clays Ltd, St Ives plc

ISBN 978–0–19–922590–3

1

For Lizzie and John

Acknowledgements

I am grateful to my editor Latha Menon for encouraging me to research and write about nothing, and to Ian Aitchison, Ben Morison, and Ken Peach for their comments that helped me to produce this something.

Contents

1

MUCH ADO ABOUT NOTHING

At some early stage in our lives most of us are suddenly hit by the question: 'Where did everything come from?' We may also wonder where our conscious self was before our birth. Can you identify your earliest memory? When I first started school, I had clear memories of the previous two or three years, especially of summer holidays at the seaside, but when I tried to recall earlier events the visions became more hazy, disappearing into nothingness. I was told that this was because I had only been born five years previously, in 1945. Meanwhile my parents spoke of a war, and of things that had happened to them before the war, but it all meant nothing to me. The world that I knew had not existed then and appeared to have been created at my birth, filled with ready-made parents and other adults. How could they have existed 'before' my conscious universe?

The weird void that was everything until 1945 continued to trouble me; then in 1969 an event occurred that was to give me a new perspective on this problem.

Apollo 10 was skimming just above the surface of the Moon, which the marvels of communications were revealing as a desolate wasteland of rock and gravel. This desert of grey dust stretched to the lunar horizon, which arced against a black void that was dotted with occasional stars, lifeless balls of hydrogen that had erupted into light. Suddenly, into this barren picture arose a beautiful blue jewel with white clouds and green continents of vegetation: for the first time humans witnessed Earthrise. There is one place at least in the universe where there is life, collections of vast numbers of atoms that have become organized such that they are self-aware and can gaze into the universe with wonder.

What if there were no intelligent life? In what sense would any of this exist if there were no life to know it? Ten billion years ago it is possible that that is how it was: a lifeless void littered with clouds of plasma and barren lumps of rock orbiting in the vastness of space. Although this epoch of 'pre-consciousness' contained no life, and must have been like some grand extension of my egocentric pre-1945 universe where gravity's dance played on without anyone being aware, nonetheless the same atoms that existed back then are what we are made of today. Once inert, complex combinations of these atoms have become organized to create what we call consciousness and are able to receive, from far across the universe, light that had set out in those earlier lifeless times. We in our 'now' can bear witness to that earlier lifeless epoch, which after the event gives it some sort of a reality. We have not been created out of nothing, but from a primeval 'ur-matter', atoms formed billions of years ago that have for a brief while been gathered into collections that think they are us.

This led to my final question: what if there were no life, no Earth, no planets, Sun, or stars, no atoms with the potential to be reorganized into future somethings; what if there were just emptiness? Having removed everything from my mental image of the universe, I tried to imagine the nothing that remained. I discovered then what philosophers have known throughout the ages: it is very hard to think about the void. As a naive child I had been wondering where the universe had been before I was born, now I was trying to imagine what there would be had I not been born at all. 'We are the lucky ones for we shall die,'[1] as there is an infinite number of possible forms of DNA all but a few billions of which will never burst into consciousness. What is the universe for the never-to-be-born or for those now dead? All cultures have created myths about those that have died, so difficult is it to accept that consciousness can just disappear when the oxygen pumps fail to power the brain, but what means consciousness for those combinations of DNA that never started, nor ever will be?

It is as hard to understand how consciousness emerges and dies as it is to comprehend how something, the stuff of the universe, erupted out of nothing. Was there a creation or was there always something? Could there even be nothing if there were no one to know there was nothing? The more I tried to understand these enigmas, the more I felt that I was at the edge of either true enlightenment or madness. Years later, having spent my life as a scientist trying to understand the universe, I have returned to such questions and taken a journey to find out what answers there are. The result is this little book. I am flattered to know that in having asked such questions, I am in good company, as in one form or another they have been asked throughout the ages by some of the greatest philosophers. Furthermore, no answer has been agreed. At various times, as one philosophy has dominated over others, the received wisdom also has evolved. Can there be a vacuum, a state of nothing? Like questions about the existence of God, it seems that the answers depend on what you mean by nothing.

In addressing these questions through the power of logic the philosophers of ancient Greece came to contrary opinions. Aristotle claimed that there could not be an empty place. This was even raised to a principle that 'Nature abhors a vacuum'; what this means and why it was believed for 2,000 years is one of the first questions that I shall address. In short summary we will see that it was not until the seventeenth century, with the emergence of the experimental method, that Galileo's students showed belief in the abhorrence of a vacuum to be due to a misinterpretation of phenomena; the apparent abhorrence was the result of 10 tonnes of atmosphere weighing down on each square metre of everything on the ground, squeezing air into every available orifice.

As we shall see, it is possible to remove the air from containers and make a vacuum. Aristotle was wrong. At least, that is the conclusion if there is only air, such that removing air has removed everything. And as science has advanced, and we have extended our senses with ever more sophisticated instruments, it has become clear that there is a lot more than just air to remove before we are left with a true void. Modern science suggests that it is impossible in principle to make a complete void, so perhaps Aristotle was not wrong after all. Nonetheless, modern scientists are happy to use the concept of the vacuum, one interpretation of modern physics being that it is focused completely on trying to understand the nature of the vacuum, of time and space in their various dimensions.

The question that I so innocently asked myself is even more enigmatic given that today we know what no one did then: the universe is expanding and has been doing so for some 14 billion years since the eponymous Big Bang. As neither the solar system, the Earth, nor the atoms that make us are expanding, the received wisdom is that it is 'Space itself' that is growing. Leaving until later the question of 'what is it expanding into', we have a further coda to my original question: if I have removed everything, then is space still expanding? In turn this begs the question of what defines space when everything is taken out. Does space exist independent of things, in the sense that if I had mentally removed all those planets, stars, and assorted pieces of matter, space would remain, or would the removal of matter do away with space as well? So let's begin our quest to see what insights wiser heads from history can offer as we try to answer questions such as: could we empty space of everything and if so, what would result? Why did the Big Bang not happen sooner? What was God doing the day before creation? Or was there always something that turned into us?

Early Ideas on No-thing

The paradox of creation from the void, of Being and Non-Being, has tantalized all recorded cultures. As early as 1,700, years BC, the Creation Hymn of the Rigveda states that

> There was neither non-existence nor existence then.
> There was neither the realm of space nor the sky which is beyond.
> What stirred? Where?[2]

Such questions were debated by the philosophers of ancient Greece. Around 600 BC, Thales denied the existence of No-thing: for Thales, something cannot emerge from No-thing, nor can things disappear into No-thing. He elevated this principle to the entire universe: the Universe cannot have come from No-thing.

The concept of No-thing was confronted with the laws of logic, Thales posing the question: does thinking about nothing make it something? The answer, according to the Greek logician, is that there can *only* be nothing if there is no one to contemplate it. My question whether there could be nothing if there was no one to know there was nothing had apparently been answered in the affirmative 3,000 years earlier, though it seems to me to have been an axiomatic assertion rather than established by argument. My quest continued but it appeared that no one after Thales defined nothing other than as an absence of something.

Having disposed of No-thing Thales then moved on to the nature of things. He successfully predicted the eclipse of the sun of 28 May 585 BC, which was a remarkable achievement and bears testimony to his ability. No wonder that his ideas were held in such high regard. He argued that if things cannot come from No-thing, there must be some all pervading essence from which all things have materialized. The question 'where did everything come from?' has inspired another: suppose that we removed everything from a region of space, would what is left be the primeval 'No-Thing'? Thales offered his solution of this mystery too: his prime suspect was water. Ice, steam, and liquid are three manifestations of water and so Thales supposed that water can take on an infinity of other forms, condensing into rocks and everything. As puddles of water seemingly disappeared, later to fall as rain from above, the idea of vaporization emerged and with it the recognition of the cycle that water provides. Space for Thales is as empty as it can be when all matter in it has been turned into its primeval form, liquid water like the ocean. Water thus contains every possible form of matter.*

* 3,000 years later this idea is defunct but modern ideas of the vacuum maintain the conceptual nomenclature by supposing it to contain an infinitely deep 'sea' of fundamental particles; Chapter 7 below.

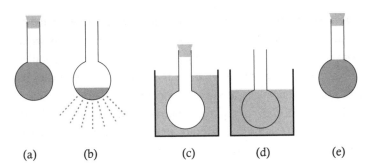

Figure 1. (a) A bottle with holes in the large bulb contains water. When the tube at the top is closed, water will stay in the bulb but if the tube is opened, (b), water will leak out of the holes. (c) The empty bottle has its tube closed and is immersed in a tank of water; no water enters. (d) Open the tube and water enters through the holes in the bulb. (e) Close the tube again and a water-filled bulb can be lifted out of the container without any liquid leaking from its bulb.

After seventy-eight years of consciousness Thales returned to the permanent void in 548 BC but the idea that there is an ubiquitous primeval essence or 'ur-matter' lived on. The nature of the ur-matter, however, was debated. On the one hand Heraclitus insisted it to be fire. So where does fire come from? Answer, it is eternal, and as such could be identified with ideas on a deity, creator of the world. By contrast Anaximenes argued that it is air. Air can be conceived of as extending infinitely, unlike water, its very ubiquity making it the preferred candidate for the universal source of all matter.

In the middle of the fifth century BC, Empedocles was faced with the question whether air was a substance or empty space. The tentative beginnings of experimental methods were brought to bear with a device known as a hydra—a glass tube, open at one end and with a spherical bulb at the other, the bulb containing holes out of which water can pour—so long as the open end of the

7

tube remains open. If you place your finger over it, no water flows. If you empty the water from the hydra and then submerge it, water will pour in and refill it so long as the open end remains open. However, if the end is covered with your finger, no water enters the holes and no air escapes either. This demonstrated that air and water coexist in the same space; no water can enter until the air leaves; air is a substance and not empty space. It would not be until the seventeenth century that Toricelli explained what was happening.

Empedocles extended the concept of ur-matter to four elements: air, water, fire, and earth. He also introduced primitive ideas on forces: for him they were love and discord, forerunners of attraction and repulsion. He was certainly the first to differentiate between matter and forces, but he still insisted that there can be no such thing as empty space.

Many forms of matter are granular. When spheres are packed together they leave spaces. So that there is no possibility of a void occurring in the 'empty' space thus created, Empedocles introduced the ether, lighter than air, which fills those spaces, indeed all space. Ether gets into everywhere, and prevents a vacuum occurring. He even imagined this ubiquitous ether being able to transmit influence from one body to another. In modern thought this is like a gravitational field.

Anaxagoras also denied the possibility of empty space and of creation of something from nothing. For him creation was order emerging from chaos rather than a material universe appearing from nothing. Order from chaos admits that things can evolve and change, as when food turns into us. This permanence of basic elements while changing their overall structure gave the idea of seeds and the birth of atomism. For Anaxagoras, there was no smallest atom, no limit to the divisibility of matter, and so no need to worry about the spaces between touching spheres, no need for gap-filling ether.

Epicurus (341–270 BC), with Leucippus and Democritus, continued the denial that something can come from nothing. They are regarded as the originators of the idea of atoms, small basic indivisible seeds common to matter. Here is born the idea that there can be a void, an empty space through which atoms move. The thinking was that if there is something already at some point, then an atom cannot move into that place; in order for motion to be possible there must be empty space into which atoms can move. They even imagined an infinite evacuated universe filled with moving atoms, which were too small to see individually but which cluster into visible macroscopic forms. Atoms are in motion but their whole is a blur, seemingly at rest. The image is like an ant hill; seen from afar it is a static mound but in close-up would be revealed to consist of millions of tiny individuals in seething motion.

Although the ideas of the atomists more nearly describe our modern picture of matter, it was Aristotle's contrarian ideas that held sway for 2,000 years. For Aristotle, a void would have to be utterly uniform and symmetric, unable to differentiate front from back, right from left, or up from down. This concept had also appeared in the Creation Hymn of the Rigveda which mused:

> Was there below?
> Was there above?

Within such a philosophy an object cannot fall or move, it can only exist in a state of rest, an idea which would eventually form a basis of Newton's mechanics. However, for Aristotle such properties denied the existence of nothing and he brought the logical arguments for the absence of a void to their clearest form. If empty space is something, and if now you place a body in this empty space, you would have two 'somethings' at the same point at the same time. If that were possible, then it would generalize to allowing any something to be in the same place as any other something, which is nonsense. So for Aristotle, logic seemed to require that empty space cannot be something and therefore is non-existent. He defined the void as where there is no body, and since the basic elements of things exist eternally, there can be no place that is completely empty.

All in all, Aristotelian logic denied the existence of the void and led to the received wisdom that nature abhors a vacuum. This was regarded as self-evident; nonetheless it was wrong, as we shall now see.

Why so Abhorrent?

The aphorism that nature abhors a vacuum was the accepted wisdom for 2,000 years, well into the Middle Ages, because it was the simplest, seemingly obvious, explanation of a whole range of everyday phenomena. Try sucking the air out of a straw: air rushes in at the other end; it is like trying to suck the air out of the whole room. So close one end by putting a finger over it and suck the air from the other end: no vacuum occurs as the straw will collapse. Or put one end of the straw in a glass of juice and suck: you end up drinking the juice. Far from creating a vacuum by sucking out the air, the liquid seemingly rises against gravity to fill the gap. It is easy, perhaps even 'natural', to think that the would-be vacuum is pulling the liquid upwards so that the vacuum cannot form. Many children do; the true answer is far from obvious. It took Galileo and some of the ablest minds of the seventeenth century to tease out the real explanation.

There are other examples that seemed to lead to the same conclusion. Place two flat wet plates on top of one another. Gently sliding one from the other is easy, but if you try to lift one it is very hard. The naive interpretation was that in doing so you would be creating a vacuum between the plates and since 'nature abhors a vacuum' it is very hard to pull them apart.

Back to the drinking straw: after sucking for a second or two, place a finger over the top of the straw while leaving the other end in the juice. A column of liquid stands proud in the tube. Release your finger and the liquid falls back into the container: so why didn't it do so when your finger was covering the top? 'Abhorring the vacuum' again. Why doesn't the liquid column split in two, the lower part falling to ground while the upper stays in the straw? The explanation was that this would require a vacuum to form at the split, at least until the lower part of the column had fallen from the tube. The survival of the liquid column was, apparently, further evidence for nature's abhorrence of forming a vacuum.

These explanations were believed for 2,000 years; they are wrong. Further confounding the discovery of the true explanation was that many regarded the abhorrence of vacuum as obvious since God would not create nothing. If in contrast you insisted that a vacuum were possible then you had to choose your words carefully to avoid running into charges of heresy. One alternative argument went something like this: God is omnipotent and so can create anything or nothing; to say that God would not create nothing is to limit God's powers; ergo a vacuum can exist. Galileo, who famously ran into such problems later, believed that a vacuum is possible, and was the first to propose testing the idea by performing experiments. This idea of testing theoretical ideas by the experimental method was radical, and also dangerous: heretics too often ended up at the stake. As a result of these experiments the reasons for the apparent abhorrence became clear and the properties of the vacuum became understood. Along the way, as understanding deepened, several instruments that we take for granted today were invented.

The Air

As children we perceive the natural order of things to be that objects in motion slow down and that light things such as paper fall to ground more slowly than stones. Galileo's experiments, which led to Newton's law that bodies continue in a state of uniform motion in a straight line unless acted upon by an external force, established the true nature of nature.

It was Galileo who first showed that air has weight. He used the fact that hot air rises and so will escape from an open container when heated. By weighing the container before and after he discovered that the escaped air had taken away some weight with it. This established that air has weight, but he could not tell its density as he did not know what volume had actually escaped. By weighing a balloon first filled with air and then with water he concluded that air is 400 times lighter than water, which given the roughness of the experiment is remarkable: the accurate value known today is a factor of about 800 at sea level.

Like anyone who has walked in a stiff breeze, he was also aware that air can exert a force, though it would be a few decades before Isaac Newton fully related force, weight, and acceleration. The air can resist motion, as when a lightweight feather is blown by a breeze and even in still air only sinks to earth slowly while a rock falls rapidly. A stone and a lump of lead, of similar size but different weight, fall at the same rate and Galileo intuitively realizes that this is the natural state of affairs: it is the air resistance that affects the feather.

The effects of air can be surprising. Its resistance to motion is why we have to keep our foot pressed on the 'accelerator' in order to keep a car moving at constant speed. The accelerator is the means of applying a force that propels the car forwards; if there were no air resistance this force would indeed accelerate the car, but the faster we move, the greater is the opposing force. It is only when the force of acceleration precisely balances the resisting force of the air that the car travels at constant speed.

The air displaced swirls around the car leaving 'thinner' air immediately behind. It is the difference between the high pressure in front and the low behind that gives the net resisting force. If the shape of the car is designed so that the air rapidly gathers immediately behind, then this pressure difference is lowered and the air resistance falls also. Designing cars, or helmets worn by racing cyclists and downhill skiers, so as to minimize the air resistance is a huge industry.

Such obvious examples were not available in the seventeenth century, which shows Galileo's genius for reducing a problem to its basics. A pebble falling through treacle comes to a stop almost immediately; in water the resistance is smaller and in air less so. He extrapolated from this and proposed that if there were no air resistance, all bodies would fall at the same rate. Although Galileo could not make a vacuum, it is clear that he had no philosophical problems with the concept of there being such a state in principle; it is just very hard to produce. This was popularly demonstrated 300 years later when the Apollo astronauts dropped a feather and a rock onto the surface of the Moon; the first demonstration experimentally appears to have been by J. Desaguliers on 24 October 1717 at the London Royal Society for Isaac Newton.

Making a Vacuum

Galileo knew that suction pumps could not raise water more than about 10 metres. Nature resists forming a vacuum but it seems there is a limit: after 10 metres of water whatever it is that is preventing the vacuum seems to be defeated. Galileo wondered what would happen if instead of water he used mercury, which is the densest liquid of all. One of Galileo's students, Evangelista Toricelli, found the answer following Galileo's advice in 1643. He demonstrated this by means of a simple experiment involving a hollow glass tube about a metre in length, sealed at one end, and a bowl filled with mercury.

A modern science textbook might describe it as follows. First use a short tube, 10 or 20 cm is long enough, and fill it with liquid. Place your finger over the open end and invert the tube, carefully lowering it into a bowl of the liquid, and don't remove your finger until the tube's open end is beneath the surface. When the open end is immersed, the liquid in the tube stays in place: a column of mercury stands proud above the surface. Toricelli did the experiment with mercury, though its toxic properties make it less popular as a demonstration today. He realized that the ability of the column to support itself had to do with the relative weights of the mercury in the tube and of the atmosphere immediately above it. More precisely, to equal the pressure exerted by the atmosphere on the mercury in the bowl, the mercury in the tube has to be a certain height.

In Toricelli's experiment this height turned out to be about 76 cm, and here was the conundrum: if a metre-long tube is filled with mercury, inverted, and then placed in the bowl of liquid, the mercury in the tube falls until the column is only 76 cm long and then comes to rest. What is in the 24 cm at the top of the tube? Where once was mercury there is now, apparently, nothing. No air could get in; Toricelli realized that he had made a vacuum.

At sea level the atmosphere weighs down on us with a force of about 1 kg on each square centimetre, which is equivalent to 10 tonnes on every square metre. A famous demonstration of how forceful air can be was made by Otto von Guerick, mayor of Magdeburg for thirty years and also a scientist with an obvious talent for popularization.

It was in 1654 that he put on his 'vacuum show' involving sixteen horses, two hollow bronze hemispheres about a metre in diameter, and the help of the local fire service. The two hemispheres were placed together to make a hollow sphere. Von Guerick showed first that it was as easy to put them together as it was to pull them apart. With the showmanship more appropriate to a conjuror, he invited members of the audience to confirm that it was easy to separate them. Now the real show began. A vacuum pump, courtesy of the Magdeburg fire department, was connected to a valve in one of the hemispheres, and the air inside was sucked out. After some minutes he announced that all the air had been removed; the valve was closed, the pump removed and the audience invited to separate the hemispheres. It was impossible. To make the point more dramatically, and it is for this that the occasion is best remembered, two teams of eight horses were harnessed together, one team being hitched to one hemisphere, and the other team to the other. Textbooks at this point simply announce that the two teams pulled in opposite

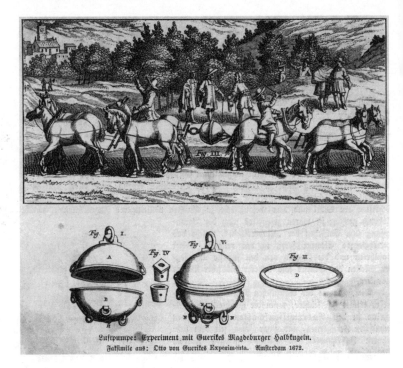

Luftpumpe: Experiment mit Guerikes Magdeburger Halbkugeln.
Faksimile aus: Otto von Guerikes Experimenta. Amsterdam 1672.

Figure 2. The Magdeburg Hemispheres.

directions and the hemispheres stayed together. The reality was more haphazard. Each of the individual horses had its own idea of what it wanted to do, and pulled this way and that. It took half a dozen attempts before von Guerick could get each team to pull in the same direction together. Finally the tug of war worked, the two teams pulled in opposite directions with all their strength, and still the hemispheres refused to part. He then opened the valve, let air back in and the two hemispheres parted easily!

In von Guerick's experiment, when the air is removed from inside the sphere, the full weight of the atmosphere is pressing on the outside with a force of 10 tonnes per square metre, with no compensating pressure on the inside fighting back. The brass was strong enough to avoid collapse but not even teams of eight horses were strong enough to provide the tonnes of force needed to overcome the external pressure.

Blaise Pascal: Water and Wine

In France Blaise Pascal was another scientist with a showman's gift. He repeated Toricelli's experiment but in place of mercury he used water and wine.

Pascal did his experiment in Rouen in front of an audience of several hundred, using tubes up to 15 metres in length, which could be raised to vertical by means of ship's masts that could be tilted. The reason for the size was because water and wine are about fifteen times less dense than mercury, whereby atmospheric pressure will support a column that is fifteen times higher, some 11 metres in all. The experiment was big scale, which was a crowd-puller, and there was a challenge: would the column of water or the wine be the taller?

You can decide; here are the two things you first need to know. Wine is less dense—lighter per litre—than water, but it is also more volatile (if you have a good nose for wine it is because you can sniff its vapour), whereas water (unless it has been strongly chlorinated) is much less so. As far as the heaviness goes, one would expect the denser water to end up lower than the wine, just as a column of mercury will be lower than both. However, what is going on in the evacuated space above the column, trapped in the top of the tube?

Realize that at the time no one believed in a vacuum—the concept of nothing was regarded as impossible. One 'explanation'

17

of what was happening was that vapour from the liquid filled the space at the top of the tube and that somehow the more volatile the liquid so the bigger the space would be. This theory would have the more volatile wine producing a bigger space and hence a lower column. However, if it is the pressure of the atmosphere forcing down on the surface surrounding the column that supports it, then the lighter wine would be taller than the column of water for the same reason that both are taller than a column of mercury.

Pascal filled the tubes, raised them so that they towered higher than the rooftops, and discovered that the wine stood taller than the water. Thus did Pascal show that volatility is not the cause of the void; it is the pressure of the atmosphere that determines the height. The space above the liquid was empty, a vacuum.*

What is a Vacuum Like?

Toricelli had made a vacuum, or at least had produced a space that contained no air, seemingly a void. But what was it: what are the properties of nothing?

In England Robert Hooke made vacuum pumps which Robert Boyle used to evacuate much larger volumes than Toricelli had been able to. This enabled him to do experiments to see what the

---

* Having just established the existence of a vacuum it's only fair to admit that the idea of vapour playing a role isn't entirely to be dismissed. There is vapour from the wine that leaks into the gap. This 'vapour pressure' pushes down slightly on the column—'slightly' because it is very tenuous compared to that of the atmosphere pushing at its base. Careful measurement of the ratio of the heights of the water and wine relative to the ratio of their respective weights would have shown a small downward push from the wine's vapour. So the space above is not totally empty, though relative to the atmosphere it is very nearly so.

18

properties of a vacuum are. He demonstrated that the air indeed disappeared by watching birds and mice being asphyxiated: a different moral philosophy operated then. A lamp could still be seen shining when viewed through a vacuum, which showed that light can travel through empty space. The sound of a bell, however, died out as the air was removed.

In France, Blaise Pascal managed to weigh the vacuum. He designed a tube with a syringe at one end and used this to suck mercury up from a bowl. The column rose and rose until it was 76 cm high, and then it stopped. Thus far was like Toricelli's experiment. Pascal now continues to pull on the plunger of the syringe. The height of the mercury stays the same but the total length of the syringe's tube grows: the amount of empty space above the mercury increases. While doing this, Pascal has the whole apparatus on a weighing machine. Throughout the entire procedure the weight stays the same. While the mercury was entering the tube this made sense as the amount of mercury was unchanged; it was just transferred from bowl to tube. Once it reached 76 cm and stopped, the increasing space atop the column grew. This was filled with 'vacuum'. Pascal had thus shown that the vacuum has no measurable weight.*

Air Pressure

A weight per unit of area is what we call pressure. On skis you can float on the snow whereas in normal shoes you might sink in:

* Actually his apparatus was not sensitive enough. In fact the weight drops as the enlarging syringe is replacing with empty space volumes that were originally occupied by air. So the true weight drops. But for Pascal's purposes the result was dramatic: whatever it is that fills Toricelli's evacuated space, it has no measurable weight.

your weight is distributed over a larger surface area in the case of the skis and so the pressure—weight per unit area—is less. The pressure of the atmosphere at sea level is the same pressure that a column of mercury 76 cm high would exert, or a column of water about 11 metres high.

If you had a column of mercury 76 cm high balanced on your head, the total pressure you would feel would be two atmospheres—one from the air and one extra from the mercury equivalent. More practical is to consider diving into the sea: salt water being marginally more dense than tap water, 10 metres is sufficient to double the pressure of the atmosphere. For every 10 metres depth, an extra pressure equivalent to that of the atmosphere is added. All of the effects that had been ascribed to the mantra that 'nature abhors a vacuum' are due to the pressure of the outside air.

As the surface area of your body is about a square metre this means that 10 tonnes of force is bearing down on you at sea level, and an extra tonne for every metre depth that you dive into the sea. So why don't you feel it? Pressure is a result of the air molecules weighing on one another. In equilibrium they push sideways, up and down the same, since otherwise there would be a net force and acceleration. This applies to pressure in fluids such as water also. The air in our lungs exerts the same pressure outwards as the external atmosphere does. Our state of comfort is the result of external pressure and internal counter-pressure in balance. A sudden change in pressure, as in a rapidly falling lift or a plane at take-off, or too sudden diving downwards while swimming, can cause discomfort. Your ears 'pop'.

A sudden change in elevation causes a change in pressure. This is because the atmosphere is finite: at high altitude the pressure is less because there is less weighing down on you as you get nearer to the 'surface'. While the sea has an abrupt surface, the atmosphere's is gradual, thinning out until eventually you reach the vacuum of outer space. This is how it was first realized.

Blaise Pascal made a seminal experiment in 1648 where he showed that the level of the fluid in a barometer depends on the elevation and from this deduced that the air pressure is critical. His brother-in-law Florin Perier measured the height of a mercury column atop the Puy-de-Dôme, 850 m above sea level, at the same time as a similar measurement was made at the bottom. The mercury column at the mountain peak was 8 cm lower than the 76 cm measured at the base of the hill. This showed that the height of the mercury column falls as the elevation rises, which is because the atmospheric pressure decreases with altitude, this in turn being because the higher you go, the less weight of air there is above you, pressing down.

Thus was invented the altimeter—a means of measuring one's altitude from the relative pressure of the remaining ocean of air above. More profound though was its implication for the nature of the atmosphere itself. It suggested that the Earth is enveloped by a shell of air that is finite; the ocean of air has a surface beyond which there is presumably nothing.* This was heretical to some religious philosophers who could not accept that God would make nugatory creations such as vacua. However, the experimental method was here exposing the failings of such superstitions, as it would in many other cases throughout the subsequent centuries.

---

* Aristotle also had thought that the air was like an ocean with a surface, but that beyond it was fire.

Today we can experience the effects of atmospheric pressure in a variety of ways. The pressure of the atmosphere drops with altitude; it is three times less at the top of Mount Everest than at sea level and mercury there would only rise 25 cm. That is how things are 10 km above us. Planes fly at such an altitude and the air in cabins has to be pressurized, typically to a level similar to that occurring naturally at about a mile high. This means that the force per square metre from the pressurized air within the cabin is much greater than that from the thinner air outside the plane. Consequently there is a force of several tonnes pushing outward on the aircraft's doors. Next time you are on a plane, notice how the doors are cleverly designed so that they cannot open directly outwards; they have first to be hauled inwards and then rotated open. The outward pressure on them actually helps retain them solidly in place during flight.

At a height of 100 km the pressure is less than a billionth of that on the ground; at 400 km a million millionth; and en route to the Moon, in space it is down by 10^{19}—an amount that is less than the size of a proton compared to a kilometre. We can thus say that essentially all of the atmosphere is in a thin shell whose thickness is less than one thousandth of the Earth's radius. Were this better known, some politicians might be more concerned about our abuse of this miraculous gas upon which we depend. As we get nearer to the top of the atmosphere, so there is less weighing down on us and the pressure drops. When astronauts fly to the Moon, they pass through more matter in the first 10 km than for the rest of their trip. Were they to travel to the furthest stars this would still hold true.

Even at ground level the pressure varies: high on a fine day and low in a storm. 'The mercury is falling' is a literally true metaphor. The idea that nature abhorred a vacuum, as religious and historical philosophers had insisted, was sent into history. As Pascal himself noted, nature does not abhor a vacuum less on the top of a mountain than in a valley, or in wet weather rather than sunshine: it is the weight of the air that causes all the phenomena that the philosophers had attributed to an 'imaginary cause'.[3]

2

HOW EMPTY IS AN ATOM?

The Electron

Electrical phenomena have been known for thousands of years, but the mysteries of the magnetic compass needle, the sparks of lightning, and the nature of electricity remained well into the nineteenth century. The situation towards the end of that century was summarized in a book that I bought as a child in a jumble sale for one penny. Entitled *Questions and Answers in Science* it had been published in 1898 and in answer to the question 'What is electricity?' it opined with Victorian melodrama that 'Electricity is an imponderable fluid whose like is a mystery to man.' What a difference a hundred years makes. Modern electronic communications and whole industries are the result of Thomson's discovery of the electron in 1897, answering the above question a full year before that book was published; news travels faster these days.

Electrons flow through wires as current and power industrial society; they travel through the labyrinths of our central nervous system and maintain our consciousness; they are fundamental constituents of the atoms of matter and their motions from one atom to another underpin chemistry, biology, and life.

The electron is a basic particle of all matter. It is the lightest particle with electric charge, stable and ubiquitous. The shapes of all solid structures are dictated by the electrons gyrating at the periphery of atoms. Electrons are in everything, so it is ironic that the discovery of this basic constituent of matter was a result of the ability developed in the nineteenth century to get rid of matter, to make a void.

For a long time there had been a growing awareness that matter has mysterious properties, although initially it did not directly touch on the question of the void. The ancient Greeks had already been aware of some of these, such as the unusual ability of amber (electron is the Greek for amber) to attract and pick up pieces of paper when rubbed with fur. In more modern imagery, brush your hair rapidly with a comb and on a dry day you might even cause sparks to fly. Glass and gems also have this magical ability to cling to things after rubbing. By the Middle Ages the courts of Europe knew that this weird attraction is shared by many substances but only after rubbing. This led William Gilbert, court physician to Elizabeth I, to propose that matter contained an 'electrick virtue' and that electricity is some 'imponderable fluid' (as in my 1898 book) that can be transferred from one substance to another by rubbing. Gaining or losing this electrick virtue was akin to the body being positively or negatively 'charged'.

Benjamin Franklin in America, taking time off from framing the constitution of what would become the USA, was fascinated by electrical phenomena, notably lightning. A thunder cloud is a natural electrostatic generator, capable of creating millions of volts and sparks that can kill. Franklin's insight was that bodies contain latent electrical power, which can be transferred from one body to another. But what this imponderable fluid was, no one knew.

Today we know that it is due to electrons, which contribute less than 1 part in 2,000 of the mass of a typical atom, and as only a small percentage of them are involved in electric current anyway, the change in mass of a body when electrically charged is so trifling as to be undetectable. How then was this imponderable fluid to be isolated, catalogued, and studied?

Electricity normally flows through things, such as wires, and as it was impossible to look inside wires, the idea developed of getting rid of the wires and looking at the sparks. Lightning showed that electric current can pass through the air and from this grew the idea that the flow of electric current might be revealed 'out in the open' away from the metal wires that more usually conduct it and hide it.

So scientists set about making sparks in gases contained in glass tubes. Air at atmospheric pressure transmitted current but obscured the flow of electrons. By gradually removing more and more of the gas, it was hoped that eventually only the electric current would remain. It was following the industrial revolution and the development of better vacuum pumps that bizarre apparitions appeared as scientists electrified the thin gas in vacuum tubes. As a result of this, electricity gradually revealed its secrets. At one fiftieth of atmospheric pressure, the current produced luminous clouds floating in the air, which convinced the English physicist William Crookes that he was producing ectoplasm, much beloved of Victorian seances, and he turned to spiritualism.

The colours of the light in these wispy apparitions depended on the gas, such as the yellow light of sodium and green of mercury familiar in modern illuminations. They are caused by the current of electrons bumping into the atoms of the gas and liberating energy from them as light. As the gas pressure dropped further the lights eventually disappeared but a subtle shimmering green colour developed on the glass surface near to the source of the current. In 1869 came the critical discovery that objects inside the tube cast shadows in the green glow, proving that there were rays in motion coming from the source of electric current and hitting the glass except when things were in the way. Crookes discovered that magnets would deflect the rays, showing that they were electrically charged, and in 1897 J. J. Thomson using both magnets and electric forces (by connecting the terminals of a battery to two metal plates inside the tube) was able to move the beam around (in effect a prototype of a television set). By adjusting the magnetic and electric forces he was able to work out the properties of the constituents of the electric current. Thus did he discover the electron, whose mass is trifling even compared to that of an atom of the lightest element, hydrogen. From the generality of his results, which cared naught for the nature of any gas left in the tube or the metal wires that brought the electric current into the vacuum tube, he inferred that electrons are electrically charged constituents of all atoms.

Once it was realized that electrons are at least 2,000 times lighter than the smallest atom, scientists understood the enigma of how electricity would flow so easily through copper wires. The existence of the electron overthrew for ever the age-old picture of atoms as the ultimate particles and revealed that atoms have a complex inner structure, electrons encircling a compact central nucleus.

Phillipe Lenard bombarded atoms with beams of electrons and found that the electrons passed through as if nothing was in their way. This almost paradoxical situation—matter that feels solid is nonetheless transparent on the atomic scale—was encapsulated by Lenard with the remark, 'the space occupied by a cubic metre of solid platinum is as empty as the space of stars beyond the Earth'.

Look at the dot at the end of this sentence. Its ink contains some 100 billion atoms of carbon. To see one of these with the naked eye, you would need to magnify the dot to be 100 metres across. While huge, this is still imaginable. But to see the atomic nucleus you would need that dot to be enlarged to 10,000 kilometres: as big as the earth from pole to pole.

The simplest atom of hydrogen can give an idea of the scales and emptiness involved. The central nucleus is a single positively charged particle known as a proton. It is the path of the electron, remote from the central proton, that defines the outer limit of the atom. Journeying out from the centre of the atom, by the time we reach the edge of the proton we have only completed one ten thousandth of the journey. Eventually we reach the remote electron, whose size also is trifling, being less than one thousandth the size of the proton, or a ten millionth that of the atom. So having made a near perfect vacuum, which led to the discovery that atomic matter contains electrons, we appear to have come full circle in finding that an atom is apparently a perfect void: 99.9999999999999 per cent empty space. Lenard's comparison hardly does the atom's emptiness justice: the density of hydrogen atoms in outer space is huge compared to the density of particulate matter within each of those atoms!

The atomic nucleus also is an ephemeral, wispy thing. Magnify a neutron or proton a thousand times and you would find that they too have a rich internal structure. Like a swarm of bees, which seen from afar appears as a dark spot whereas a close-up view shows the cloud buzzing with energy, so it is with the neutron or proton. To a low-powered image they appear like simple spots, but when viewed at high resolution they are found to be clusters of smaller particles called quarks. We had to enlarge the full stop to 100 metres to see an atom; to the diameter of the planet to see the nucleus. To reveal the quarks we would need to expand the dot out to the Moon, and then keep on going another twenty times more distance.

A quark is as small compared to a proton or neutron as either of those is relative to an atom. Between the compact central nucleus and the remote whirling electrons, atoms in particle terms are mostly empty space, and the same can be said of the innards of the atomic nucleus. In summary, the fundamental structure of the atom is beyond real imagination, and its emptiness is profound.

How Empty is an Atom?

CERN stands for the European Centre for Nuclear Research, which is what the frontiers of physics were in 1955 when it began. Today the focus of research has moved deeper, to the quarks that seed the protons and neutrons of atomic nuclei, and of many other ephemeral particles. In recognition of its mission, the laboratory is now known as the European Centre for Particle Physics. This is also more comfortable with those who regard 'nuclear' as 'nasty'. Upon approaching CERN along the road from Geneva there are the laboratory offices on one side while the fields opposite host a bizarre spherical construction, some 20 metres high and coloured dirty brown, which at first sight looks like a nuclear reactor. From afar it appears to be some disused rusting edifice, but on closer inspection is revealed to be made of wood and to bear the title: 'La Globe'.

La Globe had started life as an exhibition centre elsewhere in Switzerland. At the close, the question arose of what to do with it, whereupon rather than destroy it, La Globe was offered to CERN as an exhibition centre for its own activities. Not wishing to look a gift horse in the mouth, the CERN management accepted the offer, without a clear plan for the millions of francs that any permanent exhibition of their own would cost. One scientist proposed that this conundrum be turned to advantage: La Globe is a hollow sphere containing... nothing, so as CERN's scientists are experts on the atom, let La Globe, empty, be itself a metaphor for the atom. Even better, for a few francs a small ball, a millimetre in diameter, could be suspended at the centre of La Globe whereby visitors could 'experience' the atom's emptiness: the ball represents the nucleus, and the walls denote the outer limits of the atom. For a few more francs laser beams could play on the walls illustrating the ebb and flow of the electrons. Charge visitors an entry fee and postmodern philosophers will find contentment.

This idea was not adopted, which prevented members of the public paying money to enter a piece of art under the illusion that they were experiencing the inner emptiness of an atom. Instead temporary exhibitions, of variable relation to CERN, are housed in this blot on the landscape. But suppose that the radical suggestion had been adopted, and you had travelled across Europe with the aim of experiencing the mysteries inside the atom, paying your entrance fee, going inside the wooden ball and finding—nothing: would you demand a refund or feel that you had been exposed to a great truth?

Atoms as huge voids may be true as concerns the particles within them, but that is only half the story: their inner space is filled with electric and magnetic force fields, so powerful that they would stop you in an instant if you tried to enter. It is these forces that give solidity to matter, even while its atoms are supposedly 'empty'. As you read this, seated, you are suspended an atom's breadth above the atoms in your chair, due to these forces.

The atom is far from empty. The nucleus is the source of powerful electric fields that fill the otherwise 'empty' space within the atom. This was discovered in 1906. Rutherford had noticed that when a beam of positively charged alpha particles* passed through thin sheets of mica, they produced a fuzzy image on a photographic plate, which suggested that they were being scattered by the mica and deflected from their line of flight. This was a surprise because the alphas were moving at 15,000 km per second, or one twentieth the speed of light, and had an enormous energy for their size. Strong electric or magnetic fields could deflect the alphas a little, but nothing like as much as when they passed through a few micrometres (millionths of a metre) of mica. Rutherford calculated that the electric fields within the mica must be immensely powerful compared to anything then known. Fields of such a strength in air would cause sparks to fly and the only explanation he could think of was that these powerful electric fields must exist only within exceedingly small regions, smaller even than an atom.

From this he made his inspired guess: these intense electric fields are what hold the electrons in their atomic prisons and are capable of deflecting the swift alphas.

* Positively charged, alpha particles are composed of two protons and two neutrons. See *The New Cosmic Onion* (Taylor and Francis, 2007) or *Lucifer's Legacy* (Oxford University Press, 2000), both by the author.

31

In 1909, Rutherford assigned to Ernest Marsden, a young student, the task of discovering if any alphas were deflected through very large angles. Marsden used gold leaf rather than mica, and a scintillating screen to detect the scattered alphas. He could move the screen not only behind the gold foil, but also to the sides, and round next to the radioactive source itself. This way he could detect alphas reflected back through large angles.

To everyone's surprise Marsden discovered that 1 in 20,000 alphas bounced right back from whence they had come, to strike the screen when it was next to the source. This was an incredible result. Alpha particles, which were hardly affected at all by the strongest electrical forces then known, could be turned right round by a thin gold sheet only a few hundreds of atoms thick! No wonder that Rutherford exclaimed, 'It was as though you had fired a 15-inch shell at a piece of tissue paper and it had bounced straight back and hit you.'

After many months trying to understand these observations, Rutherford at last saw their meaning by means of a very simple calculation. The key was that he knew the energy of the incoming alphas. He also knew that each alpha particle carries a double dose of positive charge. The positive charge within the gold atoms must repel the approaching alphas, slowing them and deflecting them. The closer the alphas approach the positive charge in the atom, the more they are deflected, until in extreme cases they come to a halt and are turned round in their tracks.

Rutherford could calculate just how close to the positive charge the alphas should get, and the result astounded him. On rare occasions the alpha particles come to within one millionth of a millionth of a centimetre of the atom's centre, one ten thousandth of the atom's radius, before they are turned back. It was this that showed the positive charge to be concentrated at the very centre of the atom, leading to the picture of the 'empty' atom in terms of its particles but filled with electric field; so what is a 'field'?

Fields

Fans of Jean Michel Jarre will know his album *Champs magné-tiques*—magnetic fields. The idea of field pervades the popular consciousness with gravitational fields, or in the science fiction genre: 'warp fields in the space-time continuum'. The jargon suggests that there is a lot going on out there in the supposed void. To know what these influences are, we need first to be able to define what scientists mean by the word 'field'. It is easiest to visualize in the case where there is a definite something; so back to earth and atmospheric pressure.

A map of air pressure, which is familiar to all who worry about the weather forecast, is an example of what mathematicians know as a field, a collection of numbers that vary from point to point; in this case the numbers are the barometric pressure at each point of the country. Like a contour map, points of equal pressure can be joined by lines forming isobars: *iso* (equal) *baros* (weight or pressure).

If all that is needed to define the field is a collection of numbers, as in this case, it is known as a scalar field. The rate of change in the pressure gives rise to the winds. When isobars are far apart the breeze is gentle, whereas if they are tightly compressed, so the change in pressure is rapid, the winds are more violent. A map of wind speed is an example of what is known as a vector field. This involves both number and direction at each point, for example the speed and direction of winds (see Fig. 3).

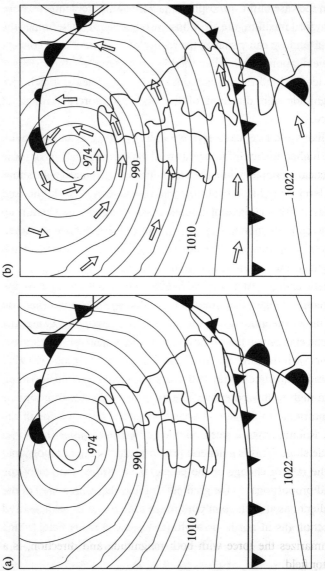

Figure 3. (a) Weather map showing pressure isobars; (b) and one also showing wind-speed vectors.

In the case of the atmospheric pressure and the winds we have a physical medium, the air, whose varying density determines the fields, and we can visualize the reality of the model. The concept of 'field' applies even when there is no material medium. This is the idea behind the gravitational and electric fields, which give magnitude and direction of the respective forces throughout space.

Hikers and mountaineers have a sense of the gravitational field. The higher up the cliff face you are, the harder you fall. That is the practical example, while the contour map showing height above sea level is the theoretical. Imagine that landscape with hills and valleys. The analogue of isobars in the weather chart is a map showing the contours of points of equal height above sea level. If you could dive into the sea unhindered, then the greater the height so the larger will be your speed of entry, the larger your 'kinetic energy'. At any initial height above sea level you have the 'potential' to gain that amount of kinetic energy; the further you fall under the influence of the gravitational force, the greater the kinetic energy you gain. The contours in the map are thus lines of points with the same potential energy, known as 'equipotentials'.

Under the force of gravity the natural motion is to fall downhill, from high to low potential. The amount of accelerating force is proportional to the rate of change of potential: the slope of the hill. Rolling down a steep hill we gather speed faster than on a gentle slope. This is a general property: the force is proportional to the rate of change of the potential, as is the strength of the wind proportional to the gradient of the isobars. So a map of the gradient has at each point both magnitude (steep or shallow) and direction (as in north- or south-facing slope). This field, which summarizes the force with both magnitude and direction, is a vector field.

Isaac Newton's insight was that falling apples and the motions of planets are all governed by gravity. The Sun is the great attractor at the centre of the solar system. If you were to fall in towards the Sun under this gravitational attraction then the further out you started, the greater will your speed be when you hit the Sun. The potential energy is thus bigger the greater your distance away from the Sun. The field of gravitational equipotentials consists of spheres with the Sun at their centre. The potential gets smaller as you move inwards, so you are accelerating from a region of high potential to one of low. The loss in potential energy is compensated by the rise in kinetic energy. This is a universal law.

The same is true if instead of massive Sun and gravity, we have electric charge and the electric field. We are all familiar with the concept of volts, even if we may be less sure of how they are defined. High voltage equates to high potential - in this case the 'potential' for inducing electric shocks, which are the result of setting electric charges into sudden motion, realized as muscle spasm. If the plates in a battery are at some positive and negative potentials, then the nearer to one another the plates are so the greater the electric field, the rate of change of potential, will be. Whereas in the case of the air we have a material medium to aid our mental image, in the case of gravitation or electric fields we do not; we have the concept and experiences of their effects but no obvious 'thing' to picture. Nonetheless their effects are measurable and gravitational and electric fields are present.

Size of Field

To get an idea of how powerful the electric fields are in atoms let's compare with what technology can do in the macroscopic world.

In a battery such as you might use in a torch or to power a radio, which will provide a few volts and for which the positively and negatively charged plates are separated by the order of a millimetre, the resulting electric field will be up to a thousand volts per metre. At SLAC, the Stanford Linear Accelerator in California, the electric fields accelerate electrons to a speed of about 300,000 km a second, within a thousandth of a per cent of the speed of light. To do this they pass through some 30 billion volts in about 3 km, which equates to electric fields of 10 million volts per metre. This sophisticated technology is giving much more powerful electric fields than in a simple battery, but is in turn nugatory compared to what it is like inside an atom. At SLAC the electric field is ten volts in each millionth of a metre; inside an atom of hydrogen, for example, some ten volts is the gap between the electron and the proton separated as they are on the average by only a tenth of a billionth of a metre. The fields within atoms are over a thousand times greater than we can achieve in macroscopic technology, though they are restricted to atomic dimensions.

The well-known rule about electric charges is that opposites attract and like charges repel. There are both types within atoms: the negatively charged electrons are at the periphery and the positive nuclear core is in the centre. When atoms are close to one another, the positively charged nucleus of one can attract the negatively charged electrons of a neighbour, causing the two atoms to move a little closer. As a result groups of atoms are mutually ensnared and clump together forming molecules and ultimately bulk matter. The most powerful electromagnetic fields that we can at present achieve macroscopically are relatively trifling compared to those within atoms because of the counterbalancing effects of positives and negatives: it is within the confines of the atom that the full power of unshielded opposite charges is realized. Once

37

this is appreciated, it is no surprise that alpha particles, even when moving at speeds of 14,000 km an hour, one twentieth that of light, could be deflected through large angles, even stopped and ejected back in their tracks: the electric fields within the atom effectively formed an impenetrable barrier.

To explore inside an atom you need to probe with something much smaller than it, which is why Rutherford used alpha particles. Far from finding a void, the invaders were repelled as if the atom is filled with a solid resisting medium. That is how the electric field manifests itself. Toricelli may have removed the air from within a region of space, but zoom in on any of the remaining atoms and you find there is definitely a 'something' in the form of the intense electric field. There is some influence throughout space caused by the presence of the electrically charged atomic nucleus. That influence remains even when all other matter has been removed.

Electric charges in motion give rise to magnetic forces whose effects can spread over vast distances as in the case of the magnetic field of the Earth. The molten metal core of our planet swirls as we rotate, the heat disrupting its atoms so that their electrons flow freely. The resulting electric currents make the Earth into a huge magnet with north and south poles, and with magnetic arms that stretch out into space. Far stronger than the Earth's gravity, its magnetic field will deflect a small compass needle. This phenomenon has been a guide for travellers and migrating birds since time immemorial. These effects were known in the seventeenth century even as the quest for a vacuum was under way. It was shown that magnetic effects and light could transmit through a vacuum, though the profound relation of light to electric and magnetic fields would not be known until the nineteenth century.

Thousands of kilometres above us, where the air is so thin as to be effectively gone, magnetic fields remain. They are critical for our existence. Cosmic rays and the solar wind of electrically charged particles are deflected by these magnetic forces. This is a crucial protective shield, as exposure to these radiations would destroy our DNA. Were the magnetic field to disappear, as is the case on Mars, it could be terminal for our species.

Pascal and Perier had shown that there is a vacuum beyond the Earth, meaning that there is no air. There is little or no gas out in space, but there is certainly a very important something in the form of the Earth's magnetic field.

Gravitational Fields and the Inverse Square Law

Gravity is the most familiar force but it is actually rather feeble: it is easy to pick up an apple, defeating the gravitational pull of the entire planet. Our muscular strength comes from the much more powerful electrical forces, which give us shape and form. However, the attractions and repulsions from positive and negative charges within matter annul one another, whereas the gravitational attraction from each and every atom in a large body adds up. Gravity rules once an object is larger than about 500 km in diameter.

Caring nothing for direction, pulling in all three dimensions the same, gravity makes spherical bodies. This is the case for the Sun, the bumps and valleys on the Earth being mere ripples on the surface caused by geological action, and its oblateness being due to its spinning around once each day.

For extremely large bodies the effects of gravity accumulate to exert a powerful pull. The Sun, no more than a thumbnail in size as viewed from the Earth, can entrap us and the planets in a cosmic waltz around the vastness of space hundreds of millions of kilometres distant. How is this influence spread throughout space?

It was Isaac Newton who had the seminal insight that gravity's pull between two bodies diminishes as the square of the distance between them increases. This 'inverse square law' of gravity's weakening with distance is critical for the structure of the universe and also possibly for the development of physical science. We are trapped on Earth that orbits the Sun; the small but relatively nearby Moon gives a gravitational tug that determines the tides, but the remote galaxies of stars don't measurably affect this. Tides, eclipses, and the flight of spacecraft can be determined without needing to take account of those distant masses. Had the force of gravity been independent of distance it would have been those remote galaxies that ruled, and the Earth would have been unable to condense under its own gravity. Had it fallen in direct proportion to distance it is possible that we could have inhabited a planetary earth but arguable whether the rules of gravity would have been determined; the ability to ignore all but two bodies, with small perturbations from a third, is what has enabled computations to be made and the basic rules to have been determined.

The inverse square law of force is not unique to gravity: the same occurs for the electrical forces between two charged particles. Given the number of possibilities that might have been, it is intriguing that both the electric and gravitational forces exhibit the same inverse square behaviour. The reason is intimately due to the three-dimensional nature of space and the fact that gravity fills all of it, as do electrical fields at least in the vicinity of a single charge.

A massive body, such as the Earth or Sun, somehow sends out its gravitational tentacles into space in all directions uniformly. The Earth's orbit around the Sun is nearly circular. Imagine the Sun at the centre of a ball whose diameter is the same as that of the Earth's orbit. The gravitational tug on our planet is the same at all points on the surface of the imaginary ball. If we now imagined ourselves transported to an orbit that was double that of the Earth's, the surface of the imaginary ball would be four times greater as the area increases with the square of the distance. Newton realized that if the force of gravity were likened to tentacles spreading out from the source in all directions symmetrically, then the intensity at any distance would be spread uniformly across the area of the imaginary ball. As the area increases with the radial distance squared, so will the intensity at any point on it correspondingly weaken.

Obviously an analogous set of remarks can be made for the electrical fields emanating from an electrically charged body.

These analogies highlight the intimate relation between the behaviour of these forces and the three-dimensional nature of space, which has been known since Newton. It gives an important clue to the mystery of how a force can occur between two apparently disconnected bodies. The intervening space is somehow involved; it is not a void but is filled with a 'field', though precisely what stuff this field consists of is a modern example of questions such as the ancient philosophers might have wrestled with. The idea came from Newton and its essential features have remained with us for 300 years, enriched by the insights of Einstein and applied in ways that Newton never knew. The basic idea is that there is a kind of tension existing in otherwise 'empty' space that manifests itself by producing forces on objects that happen to be in the vicinity. This tension's sphere

41

of influence is called a field; it is the Earth's gravitational field stretching into space that pulls skydivers to ground and the Sun's gravitational field that keeps the Earth in its annual orbit.

So an answer is beginning to emerge to the question that originally inspired me. Remove all bodies but one and its mass will give a gravitational field that spreads throughout space. This means that we could contemplate a region of space devoid of all material bodies but it would not be empty if there were even just one more body elsewhere in the universe: the gravitational field from that remote body would fill all of the otherwise 'empty' region.*

Waves

The idea of an electric or gravitational field might seem an idea dreamed up by philosophers but its reality as more than just an accounting scheme for gravitational and electrical forces can be made apparent in the form of waves. Jiggle a stick from side to side on the surface of a still pond and a wave will spread out. The motion of the stick has disturbed the molecules of water, which bump into one another, momentarily elevating some above the mean level, which then fall back down under the action of gravity, in turn pushing on their neighbours. An undulating train of peaks and troughs of diminishing intensity moves across the surface. A cork floating some distance away will start wobbling when the wave reaches it. The wave has transferred energy from the stick to the cork. More dramatic is when unstable rocks in

---

* In Chapter 6 we shall see that even that single body might not be required. According to Einstein's theory of general relativity, energy in all its forms creates gravitational fields.

the Earth's crust are suddenly displaced and fall under their own weight. Waves of compression spread through the planet and cause seismometer needles to wobble, recording the 'earthquake'. The sounds that we hear are the result of waves in the air: a sudden movement causes a wave of pressure to move outwards, and when it arrives in our ear it sets the membrane of the eardrum into motion, leading to a series of physiological responses that our brain records as sound.

In each of these cases there is a clear medium, a 'something', whose compression and dilution combined with a tendency to return to an undisturbed equilibrium creates the wave. In the case of electromagnetic waves there are analogies, and also profound differences.

If an electric charge is motionless, it is surrounded with an electric field. If it is accelerated or jiggled, an 'electromagnetic wave' is transmitted through space. An electric charge some distance away will be set in motion when the wave arrives. As was the case with the water wave or sound wave, the electromagnetic wave has transported energy from the source to the receiver. A familiar example is an oscillating charge in a radio transmitter; this generates an electromagnetic wave, which transports energy to the charges in your radio aerial.

So much for the similarities, now for a profound difference. The speed that water waves travel depends on the distance between successive peaks and troughs (the wavelength); in contrast all electromagnetic waves travel at the same speed—the speed of light. This is always true, whether you are travelling towards or away from the source. This sounds paradoxical: were you travelling away from the light source at nearly light speed yourself, you would expect that the light would only slowly overtake you; however, it rushes past at light speed itself. This bizarre phenomenon would lead Einstein to his radical new theory of space and time, special relativity, of which more in Chapter 5.

Light is a form of electromagnetic radiation, as are radio waves, microwaves, and X rays. Electric and magnetic fields fill space and can be excited into electromagnetic waves. The idea of electromagnetic waves is established fact, even if we have not yet quite come to terms with what exactly these oscillations are 'in'. Gravitational fields are also capable of giving waves, at least in theory. So what are these gravitational waves 'in'? According to theory they are ripples in space-time itself. So what is that? Is it something that remains when all else has gone? To answer that we need to start with Isaac Newton.

SPACE

Creation

It was many years ago when still a novice in popularization that I was asked to convince an Anglican bishop, versed in the creation myths of Genesis, that the universe had emerged 14 billion years previously from a Big Bang. 'Tell me; is the steady state theory no longer accepted?' God's representative asked me. The steady state hypothesis was that matter is being continually created and that, implicitly, the universe has existed for infinite time. While this avoided the great paradoxes of what was God doing the day before He made the universe, it also ran counter to observational astronomy and had fallen out of favour. I explained this and was pleasantly surprised at the reaction. The bishop almost seemed relieved: the Genesis concept was confirmed, it was just a matter of time scales.

While the bishop accepted the evidence, as do most rational people, fundamentalist 'Creationists' will argue about the time scale. As a student I first met someone who fervently and seriously believed in the 6,000-year-old universe. I explained to him the idea of parallax, how when we move from side to side nearby things appear to move relative to those further away; that the Earth's annual circling of the Sun provides enough 'side to side' motion that we can see parallax in the stars, which shows them to be light years away. Even without getting into the many other temporal measures, such as the natural radioactivity of rocks that places the Earth at 5 billion years, the evidence in front of our eyes, literally, reveals a universe that is far older than a mere 6,000 years.

He agreed, but then went on to claim what had happened 6,000 years ago was that some divine act had created a fully-fledged universe with an in-built memory: uranium in its various isotopic forms balanced so as to appear 5 billion years old; light beams created in mid-flight so as to appear to be coming from remote galaxies.

Trying to understand the universe is hard enough without adding further questions, such as if it was made 6,000 years ago, why make it with properties that suggested it to be 14 billion years old? Why did God not just start the show 14 billion years ago and let it evolve; what had God been doing for the rest of the intervening billions of years that caused Him to 'backdate' creation? Or was the universe actually created just an instant ago with each of us having an imprinted memory of our and the universe's apparent past? Such questions are not for this book as whichever way you choose to place your choice, you still have the question of what was the situation immediately before creation. Or as someone once asked me after a popular talk: 'Why didn't the Big Bang happen sooner?'

The idea of creation out of a void has plagued thinkers for as long as history has been recorded. While the ancient philosophers discussed this conundrum within the laws of logic, today we have the scientific method: experiment can test and discriminate among ideas. While science is not able to answer what happened prior to the Big Bang, or even to say whether the question is meaningful (if time itself was created at the Big Bang then what does 'before' mean?), it does suggest that there was such an event.

Ever since Edwin Hubble discovered that the galaxies of stars are rushing away from one another, the received wisdom has been that the universe is expanding. Play the vision back in time and you come to the idea that some 14 billion years ago the galaxies of stars would have been crammed on top of one another in a singular state, the outward explosion from which we call the Big Bang. Such ideas are now accepted by audiences in popular lectures but I am impressed by the range and perspicacity of some of the questions that they invite. A selection: if the universe is expanding, what is it expanding into? Are the galaxies expanding; are the atoms expanding; and when told the answer is no, then what actually is it that is expanding? If the answer is 'space', then what is that? Does space exist independent of things, in the sense that it would remain even if you took all pieces of matter away, or would the removal of matter do away with space as well?

To answer these questions we need to start with a discussion of what space actually is. This will take us from Isaac Newton and a mechanistic universe in the seventeenth century, to the remarkable insights about electricity and magnetism of Faraday and Maxwell in the nineteenth which led to the notion of space-time due to Einstein in the twentieth.

47

Newton

The classical foundations of physics, which showed how the mutual influence of one body on another results in changes in their motion, were due to Isaac Newton in the seventeenth century. His laws of motion are at first sight 'obvious' and deceptively simple. First: a physical body will stay at rest or continue to move at constant speed unless some external influence, a 'force', acts on it. This is known as his law of inertia; bodies are 'lazy' and do not want to alter their motion. To change their speed requires application of some external impetus: a force. The bigger the force so the greater the acceleration. Experience shows us that if you apply the same amount of push to a tennis ball as to a lump of lead of the same size, the tennis ball will accelerate more than the lead: Newton decreed that the relative accelerations of two bodies per unit force is a measure of their intrinsic inertia, or 'mass'. This is often referred to as Newton's second law of motion, the law of inertia being his first. Actually we see that the second law contains the first as a special case; if the force vanishes so does the acceleration and the body continues on its way undisturbed.

Every student of mechanics meets these laws and they appear to be self-evident. It is certainly true that their application enables us to send spacecraft all the way to Jupiter and by applying the right amount of force at the right time, as dictated by Newton, the craft indeed arrives at its destination. Astronomers will travel to exotic locations in order to witness the glory of a total eclipse of the Sun, their travel plans based on the faith that the predictions of Newton's laws are correct as to the exact location of the 100-km-wide band on the Earth where the orbiting Moon will be precisely in line of sight to the Sun. Newton's insights of genius are undoubtedly correct in practice, yet as soon as we start to examine them more carefully they begin to raise questions about the nature of the void.

Motion of a particle means that its position at one instant differs from that at another. Let us not get into worrying about what 'instant' or time means here, as we are about to meet problems enough anyway. What defines position? A natural and reasonable answer is, 'relative to me'. In general, the position or motion of a particle can only be defined relative to some frame of reference.

Newton envisaged some absolute space and time—a metaphorical grid of invisible measuring rods defining up–down, left–right, and front–behind: the three dimensions of space. Bodies that are at rest or in 'uniform motion' (i.e. not accelerating) relative to this moved according to his laws of motion. This grid formed the mental construct of what is known as an 'inertial frame'.

The concept goes further. Any body moving at constant velocity within this inertial frame will itself define an inertial frame. As we move, we transport our own imaginary grid of rods. Suppose I am in a car travelling at a steady 100 miles per hour along a straight road. In the frame of the car I am always positioned at the same distance from the front, in the passenger's seat suppose. In the frame of the speed camera fixed at the roadside, my position changes—in an hour I will be 100 miles away in the camera's frame, and so it duly records the fact and issues a speeding ticket.

Not all frames are inertial frames. To illustrate the idea, take a circular walk around the room. To do so you were changing direction, one moment heading north, at another eastwards. So your velocity changed: your speed may have been constant but its direction varied as you circled around. Newton tells us that a change in velocity is the result of a force acting; in this case the forces were provided by the friction between your feet and the floor, so there is no problem there. Now repeat the exercise and look all the time at some fixed point, say a chair. What you see is that, relative to you, the chair has gone on a circular tour. What forces acted on it? Gravity pulled it downwards and this was counterbalanced by the resistance of the floor, so the chair stayed still in the up–down direction. Again, no problems there. However, there were no forces acting on it in the horizontal plane, yet it appeared to go on a circular tour. This conundrum highlights an essential feature of Newton's laws of motion, and one which students frequently overlook: they apply in 'inertial frames'—frames where no net forces act on you. During your walkabout, you were being steered by the frictional forces at your feet, and so you were not in an inertial frame. The apparent circular motion of the chair relative to you violates nothing—the chair has not taken a circular tour in an inertial frame.

So what is an inertial frame? Answer: it is a frame where there are no net forces acting on me. And how do I know there are no net forces? Answer, when I am at rest or in uniform motion in an inertial frame. There is an awkward circularity in this. As we are trapped in the Earth's gravitational field, subject to its gravitational force, we are not in an inertial frame even when at rest on the Earth's surface. To make matters worse, we are orbiting the Sun subject to its gravitational whim. In practice the idea of an inertial frame is illusory. Yet in some 'common sense' manner we intuitively understand it as an approximation to an ideal, which for practical purposes enables precise computations and predictions to be made.

All is satisfactory if we imagine, as Newton did, that there is some fixed set of axes in space that defines the absolute inertial frame. Newton's philosophy of mechanics was that any two iner-tial frames must have their grids of rods moving relative to one another at constant speed (which could be zero) in a straight line without rotation. The clocks in the two frames show the same time, or at most differ from one another by a fixed unchanging amount. Thus Big Ben at rest in London, and the clock at New York's Grand Central Station, show times that differ by five hours, due to the convention of time zones, but intervals of time will be the same for both: noon to 12.20 p.m. in London equates to 7 a.m. to 7.20 in New York. If two events happen simultaneously according to a clock in one inertial frame, they will also in another. Time is thus universal and can be used by all, whatever their states of motion.

You and I, the Earth, Moon, and planets all move through this matrix without altering it in any way. The matrix is eternal, unchanging. Time behaves in a similar fashion. The tick-tick of Newton's cosmic metronome measures the passage of time as a steady flow as the bodies in the universe go through their motions.

Concepts of Space and Motion

Aristotle defined space by the bodies that it contains. He and his student Theophrastus regarded bodies as real but not space; bodies positioned relative to one another define space, but if you remove the bodies then according to Aristotle you have done away with space as well. This also implies that there can be no such thing as a vacuum as removing all the matter has removed the container—the ultimate throwing out the baby with the bathwater. Another of his followers, Strato, defined space as the 'container of all bodily objects'.[1] Strato asserted that bodies move in empty space and the container exists whether or not there is anything in it. If there is nothing in it, then it is a void.

Pierre Gassendi realized that Toricelli's experiments implied that vacua can exist and that humans can make them. He viewed space in a passive way, allowing things to pass through but not 'acting or suffering anything to happen to it'.[2]

Isaac Newton's picture of space is similar. His vision was of an absolute space, a volume in which particles, bodies, and planets exist and move. For Newton, space exists as if it were some invisible matrix of graph paper which cannot be acted on. Bodies moved through this matrix grid without altering it; its existence thus had some absolute meaning even in the absence of bodies, whereby 'empty' space is what remains when all material bodies are removed. The absence of matter implied for Newton the absence of gravitational force too, leaving nothing but the pristine inertial framework of absolute space. This is in contrast to the relative spaces defined by the grid associated with each moving particle, as these require bodies to define their relative motion and hence their relative coordinate matrices. Einstein would have none of this. He had grave doubts as to the reality of space even when there are bodies around; space and time themselves are stretched and modulated by the very motion of things. Empty space for him was an oxymoron.

In Newton's absolute space, imagine some set of events occurring such as someone juggling three balls. Now imagine the whole moving uniformly relative to this absolute space. Newton insists that this is equivalent to the same situation as before but with the matrix of space in uniform motion; the same laws and experiences apply. The Earth is moving around the Sun at a speed of some 20 km each second, so between April and October, when we are on opposite sides of the circle and moving in opposite directions, our velocity has changed by 40 km each second; nonetheless the skills for playing ball games are the same.

Whereas there is no absolute measure of velocity, only relative motions being unambiguously defined, acceleration is different: its magnitude as measured in all inertial frames is the same. A commercial boasts that a sports car can go from rest to 60 miles per hour in three seconds. This does not need any caveat 'as viewed by a stationary pedestrian in the street' as it is true also viewed by the terrified cyclist moving at a steady 15 miles per hour. However, when such commercials appear on educational television, perhaps there should be a caveat 'as perceived by observers in *inertial* frames'. For example, pedantic lawyers acting for passengers in the vehicle might dispute the claim as they are always at rest relative to the car as they speed up along with it. However, they will be feeling distinct discomfort as they are pressed back into their seats as if by some unseen force. When that car turns a sharp corner, once again the passengers will feel themselves thrust into motion, this time thrown to the side by what we call the 'centrifugal force'. In either case, the passengers are not in inertial frames.

To illustrate Newton's ideas on absolute space, and how Einstein began to question them, imagine yourself in an aircraft flying from, say, London to New York. You are sitting in the front row throughout. Just after take-off, the seat belt signs go off as the plane reaches its cruising height, whereupon it flies steadily at 500 miles per hour without turbulence for eight hours. You stayed in the same place whereas your family on the ground insist that you have travelled 4,000 miles; this shows there is no meaning to absolute position. As eight hours have elapsed, your family say that you have moved at 500 miles per hour whereas you say you have not moved at all; this shows that there is no meaning to absolute constant velocity. If to pass the time you chose to juggle balls while sitting in your seat, and assuming that such idiosyncratic behaviour did not attract the attention of nervous flight crew, the sensation and skills would be identical to what would be experienced if you were to have done this at home. Were the plane to hit turbulence, or you were to juggle during take-off, the balls would fly in new trajectories and juggling would become 'a whole new ball game'.

However, you and your family would all agree that at take-off, for maybe half a minute you accelerated along the horizontal and then were suddenly thrust skywards. Your family may have watched this, whereas you felt it manifested as a force, initially as a pressure in your back as the plane sped along the runway and then in the seat as you shot skywards. From the amount of force you could deduce, at least roughly, how much acceleration was taking place.

To Einstein, who knew nothing of jet aircraft in 1905 but could imagine being in an elevator that was in free fall, the relation between force and acceleration would turn out to be profound for his picture of space and time.

The easiest way to demonstrate acceleration without having to go too fast is on a roundabout, rotation being a particular example of acceleration.

Imagine that you are in a small windowless room on the roundabout. Isolated from the surrounding universe you can nonetheless tell that you are rotating relative to . . . something. Somehow Newton's matrix of survey rods, the space that defines space, fills your little enclosure. You cannot see it; it comes silently and cannot be heard; it has no smell and if you stretch out your arms there are no material forms to show it is there. But turn around, rotate, and you will feel it passing through your being. We call its effects 'centrifugal force' as your orientation changes. Is this absolute space therefore somehow real? Is it present even after all matter has been removed?

Ernest Mach, 200 years after Newton, proposed that the 'fixed stars' define it. We live on a roundabout as the Earth spins once every 24 hours. While your rotation on the roundabout is apparent to your gross senses, with sensitive instruments you can sense the rotation of the Earth. Even in the absence of this, a photo of the night sky with the pole star at the centre, taken with a long time exposure, will reveal the stars making circular paths around us during the night. The idea that all of those stars have made a coordinated circular dance, in some cases over millions of light years during a few hours, is nonsensical and would also require them to have travelled faster than the speed of light. It is we that have spun in an absolute way relative to the background stars.

This becomes more obvious if we speed it up.

Sit on a chair that can rotate and spin around. Everything above you, including the stars if you do this on a clear night out of doors, will spin around. What took 24 hours viewed from the rotating Earth took a mere second this time. Can the strength of your muscles send entire galaxies into motion, their speed of rotation determined by whether you make a deft push or a more forceful kick? Clearly not, and also you have no doubt that it is you and not the stars that are doing the spinning as you feel the centrifugal force. The Earth's spin also experiences centrifugal force though its manifestations are not so immediately obvious. The Earth is bulged, its diameter greater through the equator than from pole to pole; the rotations of weather systems and the tendency of motion to head 'spontaneously' to the east, known as the Coriolis effect, are other examples.[3]

Foucault's pendulum is perhaps the nicest demonstration that the fixed stars do create a frame relative to which rotations and accelerations are revealed. In many museums of science you will see a pendulum swinging from a roof from, say, north to south. Upon leaving the museum some hours later, the pendulum is swinging east to west without anyone having intervened to change its direction. This phenomenon at London's Science Museum was one of the fascinating mysteries that excited me into science as a child. The answer is that indeed nothing has given it a push; the pendulum is still swinging in the same direction *relative to the fixed stars*; it is the Earth beneath that has spun taking the museum and us with it.

While this view of the stars makes sense of the idea that rotation is indeed absolute, can it really be that a child innocently riding on a roundabout is sensing that there are galaxies of stars remotely distant? The Earth is so immediate that its gravity holds us to ground. The Moon is small but near enough to affect the tides while the nearest star, the Sun, holds us in our annual orbit. The gravitational pull of other planets is so small as to be unmeasurable, whatever advocates of astrology might claim. Remote stars and even individual galaxies of billions of stars have too small a gravitational pull to affect our daily affairs, though the collective gravity of our Milky Way and that of the Large Magellanic Cloud galaxy holds the latter as a satellite around us. As one doubles the distance away from here, the gravitational effect of any galaxies will die to a quarter. However, if the galaxies are on the average uniformly spread throughout space, then each doubling of distance will quadruple their number with the result that their total gravitational effects stay roughly the same all the way to the far reaches of the cosmos. So although our daily tides and annual seasons are controlled by the gravitational pulls of Sun and Moon on the wobbling Earth, this is all acted out in the background gravitational field of the distant 'fixed' stars.

This is the nearest we can easily come to identifying the absolute metric grid of absolute space. Rotations relative to this gravitational matrix are what we feel as we spin on the round-about, turn corners in a car, or in general change our velocity. Problems arise however; the fixed stars are not so fixed; this picture would also imply that the night sky should be as bright as daytime since there should be a star in every part of our field of view (known as Olbers's paradox) and a resolution of this is that the universe is expanding. The picture of space and time that we have presented, based on the philosophy of Isaac Newton, is known to be incomplete. Since the start of the twentieth century the richer picture of Albert Einstein has ruled. The origins of this go not to gravity but to electric and magnetic effects, though gravity will turn out to play a major role.

4

WAVES IN WHAT?

Electromagnetic Fields and Waves

The next time that you turn the key to start your car and the current from the battery magnetically stirs the motor to life, give a moment's thought that in what just happened lie the seeds of relativity and the modern view of space and time. When Michael Faraday experimented with electricity and magnetism early in the nineteenth century, no one anticipated that this would lead to a profound re-evaluation of Newton's world view. He made discoveries of such magnitude that had the Nobel Prizes existed in the nineteenth century, Faraday could have won as many as six, the most far-reaching of the discoveries being that electric and magnetic fields are profoundly interwoven and affect one another.

Examples of this are what happen in your car engine. For example, move a magnet suddenly and you will create *electric* forces; this is known as induction and is the principle behind electric generators. The idea is that variable magnetic fields give rise to electric fields and vice versa: sudden changes in electric fields give rise to magnetism. The swirling electric currents within the centre of the Earth give rise to its magnetic field. The ability of electric and magnetic fields to ebb and flow is an integral part of the electric motor.

Magnetic fields can be created by electric currents, which in turn are electric charges in motion. So far so good, at least until you ponder 'in motion relative to what?', to which a reasonable answer is 'relative to you' in your (static) inertial frame. However, suppose now that you moved alongside the wire carrying the current, and at the same velocity as the flow of electric charges within it. In such a case you would now perceive the charges to be at rest. An electric charge at rest relative to you, in an inertial frame, gives rise to an electric field, so in this situation you perceive there to be an electric field whereas previously you felt magnetism. Speed up or slow down and magnetic fields will emerge at the expense of the electric ones. What was a magnetic field in one inertial frame has become an electric field in another. Whether you interpret the field as electric or magnetic depends on your own motion.

Einstein insisted that the laws of physics cannot depend upon the uniform motion of the observer. What is good for one observer in one inertial frame must be so for all in all inertial frames whatever their relative motions. This led to his theory of relativity, of which we will see more in Chapter 5. What it did for electricity and magnetism was to show that they are not separate and independent sets of phenomena but that instead electric and magnetic fields are profoundly intertwined into what is known as the electromagnetic field.

This provided the basis for the theory of electromagnetic phenomena that had been created by James Clerk Maxwell in the mid-nineteenth century. Maxwell had encoded Faraday's discoveries and all known electric and magnetic phenomena into just four equations. Having formulated them, he then worked out their solutions and in doing so he discovered that they implied a whole symphony of new phenomena.

To understand what and why, first realize what Maxwell's equations were designed for. They summarized that a changing electric or magnetic field would generate its complementary partner: electric to magnetic and vice versa. An electric field is a vector field: it has both magnitude and a direction. If the electric field was oscillating, such that the directions 'uphill' and 'downhill' were interchanged N times each second, the resulting magnetic field would also oscillate at the same frequency. That is what his equations implied. Next he put the case of the oscillating magnetic field into another of his equations and he found that it predicted that this would produce a pulsating electric field. Put this electric oscillation back into the original equation and you find that the sequence goes on, electric to magnetic, back and forth. The resulting effect is that the whole mélange of electric and magnetic fields would propagate across space as a wave. Faraday's measurements of electric and magnetic phenomena provided the essential data that, when inserted into Maxwell's equations, enabled the speed of the waves to be calculated. Maxwell found this to be 300,000 km each second, independent of the frequency of the oscillations. This is also the speed of light, from which he made the seminal leap: light is an electromagnetic wave.

Visible light, the rainbow of colours, consists of electromagnetic waves whose electric and magnetic fields oscillate hundreds of millions of times each second, the distance between successive crests in intensity being in a narrow range around a millionth of a metre. What we perceive as different colours is the result of electromagnetic waves oscillating at different frequencies. Maxwell's insight implied that there have to be other electromagnetic waves beyond the rainbow, travelling at the same speed as light but oscillating with different frequencies.

Infrared and ultraviolet rays were already known, the 'infra' and 'ultra' referring to their oscillation frequencies relative to visible light. These clues inspired scientists to look for other examples. Heinrich Hertz in Karlsruhe produced electric sparks and showed that they sent electromagnetic waves across space without the need for material conductors. This was the original source of the name 'wireless'. These primitive radio waves are electromagnetic waves akin to light but in a different part of the spectrum. Hertz has given his name to the unit of frequency such that once a second is known as one hertz, and thousands or millions of times per second are kilohertz and megahertz. Radio waves are electromagnetic waves that are oscillating in the kilohertz to megahertz range.

Just as was the case for visible light, which travels through a vacuum, so it is for radio waves and all frequencies of electromagnetic waves. We can communicate with remote spacecraft courtesy of radio waves. They travel through empty space as do the rays of visible light, and at the same universal speed of 300,000 km per second.

Another implication of Maxwell's work was that electrically charged bodies and magnets separated by large distances do not interact with one another instantaneously but do so by means of the electromagnetic field, which spreads out from one to another body at the speed of light. Jiggle an electric charge at one location, and it is only when the resulting electromagnetic wave reaches a remote charge that the latter will start oscillating in concert. This was utterly different from the mechanical picture of Newton where such action occurs instantaneously.

Radio reception, X-ray crystallography, and seeing in general involve the ability of an electromagnetic wave to be absorbed or to be scattered by matter after having passed through seemingly empty space. Here lies the basic question about light as an electromagnetic wave: what medium is waving, or as it was posed more bluntly, 'waves in what?'

Waves in What?

In the seventeenth century Robert Hooke had discovered that sound does not pass through a vacuum. This made sense as the fact that sound is simply vibrations of the air had been known since the Stoic philosophers in ancient Greece; take away the air and the sound goes too. This contrasted with light and magnetism as a lamp burned as brightly when viewed though the vacuum as though air and magnets in a vacuum continued to influence one another. So after the air had gone, did something else remain that was capable of transmitting these effects? The ancient Greeks had so abhorred the notion of the vacuum that the idea of the 'ether' had grown—a 'medium subtler than air' which filled all of space even after air had been removed. Isaac Newton believed in ether though precisely what he took it to be is not clear. Ideas of the ether abounded in the subsequent centuries, until finally overturned by Einstein's theory of relativity. This is how the ether came and went.

Newton was a mechanical philosopher, explaining natural phenomena as the motion of particles in matter, which led him initially to picture light as a burst of corpuscles or, as we now call them, 'photons'. Newton's mechanics also denied the idea of 'action at a distance'. The phenomenon of electrostatic attraction, such as when a scrap of paper is attracted to a piece of glass that has been rubbed by cloth, he described as due to some ethereal substance that flows out of the glass and pulls the paper back with it. In 1675 he produced his theory of light, which contained a universal ether.

But he was not happy. Within five years he had given up on the ether and introduced the idea of attractions and repulsions between particles of matter. Thirty-five years later he produced the second edition of his treatise *Opticks* in which he again accepted the existence of an ether, but one that allowed action at a distance by means of repulsion between the particles that comprised the ether.

In the eighteenth century the Swiss mathematician and physicist Leonhard Euler rejected Newton's corpuscular theory of light and gave his own explaination of optical phenomena as vibrations in a fluid ether. Everything changed at the start of the nineteenth century when the English physician Thomas Young showed that light consists of waves. His primary interest was in perception. As a medical student he had discovered how the lens of the eye changes shape as it focuses on objects at different distances. He discovered the causes of astigmatism in 1801 and then became interested in the nature of light. It was from this that he discovered interference effects, where light passing through two pinholes would give rise to series of alternating dark and bright bands. This was analogous to the way that waves of water can mingle, giving large peaks where two crests coincide or flatness when a trough and a crest meet. The analogous mingling of peaks and troughs in the undulating waves of light would naturally explain this phenomenon; indeed the idea that two pieces of light could create darkness was remarkable and its explanation in terms of waves was taken as a definitive proof of the wave nature of light*.

Interest in the nature of light and electricity in the nineteenth century led to resurgence of the old idea of the ether as the medium that transmitted light waves much as air transmitted sound. This ether in nineteenth-century science was postulated to be weightless, transparent, frictionless, in effect undetectable by any physical or chemical process. It permeated everything and everywhere, supposedly some form of elastic solid, like steel, yet with the remarkable ability to let the planets pass through as if it were not there. Much of nineteenth-century science was taken up with trying to detect this mysterious stuff.

The ether idea solved the conundrum of transmission through a vacuum but did not explain why light changed its behaviour when passing through transparent media that were definitely not vacuum, such as water or glass. The speed of light through water is slower than in a vacuum; some materials that are transparent to light viewed directly nonetheless can become opaque to light that has been scattered along the way, a phenomenon that is exploited in some polarized sunglasses. All these phenomena were explained naturally following Maxwell's proposal that light is a wave of electric and magnetic fields.

Ether was supposed to be the medium in which light oscillated. The assumption was that the ether was stationary throughout the universe, defining Newton's absolute state of rest. By 1887 it was becoming clear that light is a wave of oscillating electric and magnetic fields. In the case of sound the wave is in the

---

* Except in England where opposition to Newton's theories was regarded negatively. Young's work became accepted after it was reproduced by the French physicist Augustin Fresnel.

direction of motion; the electromagnetic wave is different in that the oscillations are at right angles to the motion. The laws of electromagnetism and of light were assumed therefore to apply to this ideal case of situations relative to a static ether.

The Problem of the Ether

Maxwell's calculation of the speed of electromagnetic waves offered a way of measuring our velocity relative to the ether that defined absolute space. For illustration think of water waves. Drop a stone into water and a wave spreads out. The speed of the wave is about a metre per second. This speed is a property of the water; it does not depend on the velocity of the source. If the stone is dropped in from a stationary boat, the waves spread at 1 metre each second; if dropped in from a speedboat they still spread at 1 metre per second. If you are on a boat that is at rest in the water, you will see the waves pass you at a speed of 1 metre each second. If however you were heading into the waves at 10 metres per second the waves would approach you at 11 metres per second, whereas if you were headed the other way at the same speed relative to the water, you would be overtaking the waves at 9 metres per second. You can determine your absolute speed relative to the water this way.

As it was for the boat in the water, so it would be for the Earth in the ether. Electromagnetic waves move at 300,000 km each second, which is a property of space and independent of the speed of the source as in the water example. And by analogy with that, if we were to move through the ether, by measuring the speed of the light waves we would be able to determine our velocity relative to that medium. All that was needed was to measure the velocity of light through the ether in a variety of directions and from this it would be possible to determine in which frame the speed was exactly that calculated by Maxwell. This frame would then be the absolute frame of the cosmos: the state of absolute rest relative to the ether. However, things did not turn out as expected.

Newton's laws of motion were assumed to apply (no alternative existed!), so if the Earth moved relative to the ether then it ought to be possible to detect its motion. For example, if the Earth is travelling through the ether, then light travelling in the same direction would have a speed elevated by that of the Earth, whereas that at right angles would not get this extra boost. Only when at rest relative to the ether would the 300,000 km per second speed of light, as calculated in Maxwell's theory, arise.

The Earth is about 150 million km from the Sun and completes its annual circle of some billion kilometres in a year, 30 million seconds, which implies that the Earth travels some 30 km each second. According to Maxwell, light travels at 300,000 km per second 'relative to the ether' and so the motion of the Earth when at two points diametrically opposed in its orbit will have changed its velocity relative to light by about 1 part in 5,000.

Detecting such subtle effects from the Earth's motion required some ingenuity. Albert Michelson made a first attempt in 1881, but it was not until 1887 in collaboration with Edward Morley that the required accuracy was achieved. They did this not by comparing measurements six months apart but in a single experiment in the laboratory where they split a beam of light in two, and sent them in different directions, finally bringing them back to the starting point by means of mirrors. The two beams travelled perpendicular to one another, so if one happened to be parallel to the Earth's motion the other would be perpendicular to it. The two beams would be affected by the ether in different ways and their return to the starting point after reflection would occur at slightly different times. As the electromagnetic waves oscillate at some frequency, the slight time difference will be manifested as a difference in the amplitudes of their wobbles.

If the light waves had fallen out of step, in the sense that one had oscillated slightly more times than the other as a result of its speed being affected by the ether, the peaks and troughs of the two beams would lead to interference bands, dark and light. By measuring their widths and number, it was possible to make extremely sensitive measurement of the relative velocity of light travelling in the two perpendicular directions. Initially Michelson did the experiments in Berlin alone and then later did more precise ones in the USA in collaboration with Edward Morley. No interference fringes were seen and the conclusion was that the Earth is not moving through the ether or, as Michelson stated more precisely: 'the hypothesis of a stationary ether is erroneous.'

This conclusion is logically correct; the full implication allowed different immediate possibilities. One is that the ether is like the story of the king who was supposedly wearing fine clothes that could only be seen by the wise while appearing invisible to fools. Of course everyone claimed to see the quality of his dress, until a child, who had not been told of the tale, announced correctly that the king was naked. The analogy here is that there is no ether at all. This is now the received wisdom following Einstein's work, of which more later. The second possibility was that the Earth is dragging the ether along with it by friction. Our motion through the ether creates huge eddies, so that the light beams move through an ether that is stationary relative to the laboratory even as we all move through the more remote streamlined ether.

Newton recognized that bodies feel resistance even when moving through the air. The state of permanent unaltered movement that formed the basis of his laws of motion was to be achieved by removing all such impediments, and hence the ether had to be completely tenuous. For example, the motions of the planets were described by Newton's laws and their success implied that the planets move freely in the gravitational influence of the Sun; there is no room for interacting with an ether here. However, this created an immediate paradox, since if the Earth is to drag the ether around it, it must be interacting with it and then the success of Newton's mechanics applied to the planets becomes a problem. Nonetheless imaginative proposals were made. George Stokes (1819–1903) was a British physicist famous for his studies of viscous fluids. He believed in the wave theory of light and also in the ether, which he suggested behaved like wax, being rigid but also being able to flow when subjected to a force. This led some to suggest that the motion of the planets supplied the force that made the ether flow, and that they dragged the ether along with them by friction, but no experimental evidence for this was ever found. That, together with the ad hoc nature of the proposal, led to its downfall.

There was a remarkable third alternative interpretation discovered by George Fitzgerald in England and Hendrik Lorentz in Holland. Independently they noticed that if bodies moving through an ether contracted in their direction of motion by an amount that depended on the square of the ratio of the velocity of the Earth to that of light, then the motion through the ether would be masked and the results of Michelson and Morley could be explained.

The idea was as follows. Suppose we have a metre rod at rest on earth. Now imagine that rod moving rapidly past you, through the ether. Lorentz and Fitzgerald supposed, correctly, that the forces holding solids such as rods together are electromagnetic, and that motion through the ether disturbs them. Using Maxwell's theory they calculated that at velocity v relative to that of light, c, the rod would contract in length by a fractional amount

$$\sqrt{1 - \frac{v^2}{c^2}} \qquad (1)$$

At Earth speed of 30 km/sec the effect is less than one part in a million: a metre rod would shorten by about a micron.

In this theory the apparatus that Michelson and Morley were using would shrink as it headed into the ether whereas when moving at right angles to the ether its length would be unaffected. This subtle difference in distance between the contracted lengthwise and unaltered perpendiculars precisely matches the expected time delay whereby the two beams return in step, consistent with the result of their experiment. So in this explanation, space can be filled with ether and it is inherently impossible to detect motion through it as the measuring instruments conspire to hide it!

Their explanation also implied that motion through the ether would modify the resistance to acceleration, the inertia or mass of the moving body increasing in proportion to $1/(1 - \frac{v^2}{c^2})^{1/2}$. Thus as the velocity approaches that of light, $v = c$, the mass would become infinite. As a consequence an infinite amount of force would then be required for any massive body to reach light speed. Although the idea seemed artificial and was not widely accepted as an explanation, in 1901 electrons emitted in radioactive beta-decays, with a variety of speeds, were found to have a mass that varied with velocity in agreement with this formula. This made people take note of the Lorentz–Fitzgerald transformation, as it became known.

Today we know that these velocity-dependent transformations are correct. Lengths do contract and masses do grow with increasing speed in proportion to the factor $1/(1 - \frac{v^2}{c^2})^{1/2}$ but not for the reasons that Lorentz and Fitzgerald had suggested. Einstein took a new perspective on the problem. The invariance of the speed of light with respect to the speed of source or observer is a result, in part, of distances contracting as in Lorentz and Fitzgerald's formula but this was not due to any ether acting on the rod. For Einstein the contractions are an intrinsic property of space itself. Distances and time intervals as recorded by observers at different speeds take on different measures; what is space for one observer is a mix of space and time for another. These ideas, which are the basis of Einstein's theory of relativity, formed a completely new world view.

TRAVELLING ON A LIGHT BEAM

Michelson and Morley's experiment showed that the Earth does not measurably move relative to the ether. Lorentz and Fitzgerald had proposed that the ether distorts the measuring apparatus so as to mask the motion, but Albert Einstein realized that there was a more radical explanation: there is no ether at all!

The fact that the velocity of light is independent of the speed both of the source and of the receiver was an enigma, though it is not clear to what extent Einstein was aware of this result.* In any event he had begun to muse about the symmetry of things with respect to motion. If there is no ether there is no absolute space and hence no absolute motion: only relative motion has physical meaning.

Einstein knew that light is electromagnetic radiation whose properties are described by Maxwell's equations. He thought

---

* See page 74 and also A. Hey and P. Walters, *Einstein's Mirror* (Cambridge University Press, 1997), 50.

about how this radiation would appear to two observers who were moving relative to one another. Specifically he made a series of 'thought experiments', more usually referred to by their German analogue 'gedankenexperiment', which involves imagining a situation according to the laws of physics.

At the age of 16 Einstein had wondered what it would be like if he travelled on a light beam. If light had been electric and magnetic vibrations in the ether similar to sound waves being vibrations in the air, then the analogy would be that as sound travels at Mach 1 relative to the air, so light travels at 300,000 km per second relative to the ether.* There were no jet aircraft in 1900 but had there been he might have imagined one at Mach 1, keeping pace with the speed of sound, flying at the same speed as the pressure waves that are propagating through the air. If now we replace air by ether and sound by light, we could imagine travelling along with the light wave. This had bizarre consequences if the analogy with sound were correct. First, if you looked in a mirror your image would have vanished: the light from you is heading towards the mirror at the same speed as you are and so will not arrive at the mirror, let alone be reflected, until you get there. This was psychologically weird, but as far as I can tell, there is nothing that says one's image is so sacrosanct that this could not have been the outcome. Where the physical inconsistency emerged was when he considered what Maxwell's theory would allow. If you pursued and eventually caught up with an oscillating wave of electric and magnetic fields, and then travelled alongside them at speed c, you would perceive an electromagnetic field that is oscillating in space from side to side but not moving forwards, remaining at rest. Yet there was no such thing in Maxwell's equations:

---

* This speed of light is traditionally denoted by the symbol c. Think from here on that the symbol c, stands for 'c'onstant light speed.

oscillating electromagnetic fields move at speed c. Seemingly if Maxwell's theory of electromagnetic phenomena was correct, and everything we know says that it is, then the situation that Einstein was imagining, travelling at the speed of light, must be impossible: we can never achieve light speed.

This set Einstein thinking about the definition of velocity, and the concepts of absolute and relative. For this gedankenexperiment he imagined a passenger on a train watching another train pass by, which was inspired by a phenomenon that we have all at some time experienced.

You are sitting in a train that has stopped at a station and on the adjacent line is another train, which has also temporarily stopped, but headed in the opposite direction. Impatient to depart you notice that at last you are moving relative to the carriages of the adjacent train, and so gently as not to have been aware of the slight force of acceleration. Then as you pass the final carriage, you discover that you are still at rest in the station and it is the other train that has departed. As Einstein is supposed to have asked when travelling from London during his time at Christ Church College, Oxford, in the 1930s: 'what time does Oxford arrive at this train?'* In these examples there is a concept of absolute rest, namely the situation of the station and surrounding landscape. Einstein argued that if this experiment were performed with two trains, each moving at constant speed in a vacuum with no ether to define an absolute state of rest, there would be no measurement that could tell which one was moving and which was at rest. Maxwell's equations describing the behaviour of electric and magnetic fields would therefore have identical consequences for

* This story is often apocryphally attributed to Einstein and Cambridge, but I and my publishers are in Oxford as is Einstein, at least as seen by aliens 75 light years away.

two trains, and in particular the speed of light also would appear to be the same for both.

Michelson and Morley had established this phenomenon experimentally, though whether Einstein was aware of this or deduced that the speed of light is constant by means of the above thought experiment is argued. At various times Einstein claimed that he was unaware of the experiment in 1905 when he created his theory of special relativity. However, in 1952 he told Abraham Pais[1] that he had been conscious of it before 1905, through his reading of papers by Lorentz, and 'he had assumed this result of Michelson to be true'. No matter; the phenomenon is there to be puzzled over as it runs counter to intuition unless 'common-sense' ideas about space and time, as articulated and accepted since Newton's time, are wrong.

Space, Time, and Space-time

Speed is a measure of the distance travelled in an interval of time. According to 'common sense' or, better, to Isaac Newton, the metre-sticks and chronometers that measure space and time are the same for all. Speed is the ratio of distance moved to time elapsed and relative speeds add or subtract depending on whether you are heading towards or are running away from a speeding object. However, common sense fails for light beams, since independent of how fast you move or in what direction, your relative speed to a light beam is invariant. Einstein realized that something must be wrong with our concepts of space and time.

What does simultaneity mean? If two things happen 'at the same time' for someone on Earth and for an astronaut on Mars, how are they to know that their clocks are synchronized? If we could send a time signal to Mars instantaneously then all would be well, but in reality there is a time delay as it can only travel there at light speed, c. Upon receiving our signal the Martian could send us a signal acknowledging receipt, and we could then adjust our clocks accordingly. It seems straightforward. However, the planets are in motion; at first sight that also can be taken into account but another of Einstein's gedankenexperiments reveals that, metaphorically at least, there is more to this than meets the eye.

Einstein must have loved trains. Imagine you are at the middle of a stationary train and send a light signal to the driver at the front and the guard at the rear. They will receive the signals at the same instant. This fact will be agreed upon by you and also by someone standing at the side of the track adjacent to you. Now suppose that instead of being at rest the train is moving at a constant speed. I am at the trackside and as you pass me you send the light signals to the driver and the guard. You will perceive them to arrive simultaneously but I will not, because the light does not get there instantaneously; in the brief moment that it took the light beam to travel from the middle to the ends of the train, the front of the train will have moved away from me while the rear will have approached. From my perspective the signal will reach the guard a few nanoseconds* before it reaches the driver, whereas you will insist that they arrived simultaneously. Simultaneity as recorded by someone on the train is not simultaneity for someone on the trackside; our definition of time intervals, the passage of time, depends on our relative motion.

---

* A nanosecond is a billionth of a second. In a nanosecond a light beam travels 30 cm or, in old units, a foot, which is about the size of your foot.

Einstein had realized that the bizarre fact that c is a constant, independent of the motion of the receiver or transmitter, somehow links with the notion of time differing for people who are in motion relative to one another. There is one 'natural' way that c could always be the same, that is if it were infinite, in which case signals transmit instantaneously and we don't need to think much about what is involved in defining, measuring, and comparing time intervals. In our daily experience c is all but infinite and these subtle properties of time are not noticed. Einstein's insight was that c being *finite* and constant means we must think again about things that we have previously regarded as obvious, if we have ever thought about them at all.

He worked out the logical consequences and found that not just time intervals, but also distances, as measured in one inertial frame differ from those in another, the mismatch depending on the relative speed between the two inertial frames. Spatial intervals shrink and time is stretched by a common amount; for two people moving at speed v relative to one another the ratio is $\sqrt{(1 - \frac{v^2}{c^2})}$, which is the same factor that Lorentz and Fitzgerald had introduced in their ether theory. At the speeds that we are used to in normal experience, this factor is so near to one that these surprises are not noticed, but for fast-moving atomic particles, such as occur in cosmic rays or in accelerator laboratories such as CERN, the effects of relativity are critical.

Common in cosmic rays is a particle called the muon. If you were a newborn muon, you would expect to have but a millionth of a second to live. Created by cosmic rays at the top of the atmosphere, 20 km above the ground, and moving nearly at 300 million metres each second, you will only be able to travel at most some 300 metres in your lifetime. What is remarkable, then, is that you manage to reach the ground; muons from cosmic rays are passing through this page right now. How can you travel 20 km from the upper atmosphere in a millionth of a second? You cannot. The explanation is that time and space, perceived by the flighty muon, have a different beat from that of the observer on the ground.

A millionth of a second elapses for the clock within the muon, but moving at high speed towards ground $1/\sqrt{(1 - \frac{v^2}{c^2})}$ might be several thousand. The time as measured by a clock on the ground will be stretched to hundredths of a second, which is plenty enough time to travel 20 km.

So the paradox is explained from the space-time matrix that the ground-based observer lives in, but how does it work from the perspective of the muon? Inside the muon, your egocentric perspective regards yourself as being at rest and the ground to be rushing up towards you. What the ground-based observer measures as a 20-km-high atmosphere, you see shrunk to a fraction $\sqrt{(1 - \frac{v^2}{c^2})}$, which makes the distance appear to be only a few metres. Your clock says you have a millionth of a second in your time to travel but a few metres in your space; once again, no problem.

This contraction of length precisely correlates with the dilation of time so that the speed of light, whose dimensions are a ratio of distance and time, is the same for moving muons as for stationary scientists on the ground. Had light travelled at infinite speed so that signals could be sent instantaneously, none of these 'unnatural' things would have worried us and intervals of lengths or time would be measured the same by everyone. In such a case muons also could have travelled up to infinite speed, fast enough to get to ground instantaneously. It is the fact that c is finite that makes the structure of space and time depend upon our speed; it is because

77

c is finite but *large* that we normally are unaware of this in our sluggardly affairs.

Although space and time intervals are changed from frame to frame, Einstein's analysis revealed that a particular combination stays the same. This gives our modern picture of space-time. The invariant combination can be illustrated by a concept familiar in geometry in two or three dimensions but extended to four dimensions—three of space with time treated as a fourth dimension.

Space-time

It was Maxwell's theory of electric and magnetic phenomena with electromagnetic radiation having a universal speed that had led to Einstein's new world view as expressed in his Special Theory of Relativity. Einstein had showed how Newton's concept of inertial frames with its metaphorical grid of measuring rods and constant flow of time is but an approximation of a more profound picture. The German mathematician Hermann Minkowski then noticed that this theory took on a familiar form if space and time are intertwined in what has become known as four-dimensional space-time. We are all familiar with Pythagoras' theorem for a right-angled triangle in two dimensions, such that if x and y are the distances along the horizontal and vertical sides, then the square of the distance along the hypotenuse is the sum of the individual squares: $s^2 = x^2 + y^2$. We could think of x and y being the latitude and longitude of some point; we could rotate the map giving new lines of longitude and latitude, rotated relative to what we had previously but still being perpendicular to each other. The magnitude of s^2 would remain the same in terms of these new x

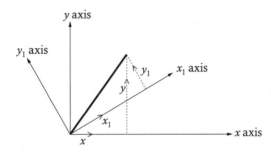

Figure 4. Pythagoras tells us that $s^2 = x^2 + y^2$ independent of which perpendicular direction x and y are.

and y; we say that s^2 is invariant under rotation (see Fig. 4). In three dimensions we have latitude, longitude, and height z above the ground. The invariant distance measure then generalizes to $s^2 = x^2 + y^2 + z^2$. This is true in any inertial frame: whether rotated or displaced, the distance s^2 remains the same.

In special relativity as one goes from one frame to another, this quantity is not invariant. In a fast-moving frame the distances have shrunk. However, the time clocks also tick at a different rate. It turns out that the following combinations of space and time in two frames remain invariant: $s^2 = x^2 + y^2 + z^2 - c^2t^2$ where c is the (invariant) speed of light. Minkowski proposed that space and time be thought of as a single four-dimensional space-time; the subtly different aspect of the dimension of time versus space being in part encompassed by the minus sign that appears in front of the time coordinate, in contrast to the positive signs for the spatial ones.

All of these insights had come from thinking about how electromagnetic phenomena appear to observers in different inertial frames (Einstein's paper was titled 'On the Electrodynamics of Moving Bodies'). Nowhere had gravitational effects been

79

included in the thought experiments. Einstein realized that this made his theory incomplete. The electron and proton inside a hydrogen atom interact through the electromagnetic force and also through gravitation; the space-time fabric must be the same for both interactions otherwise which would rule? In general it was inconceivable that the void in which electromagnetic phenomena take place operates with a different space-time fabric from that which applies for the all-pervasive force of gravity.

According to Newton's theory of gravitation, the Sun and Earth interact instantaneously. However, according to Einstein's relativity theory of 1905, such interactions can only be transmitted at the speed of light, such that time would elapse in gravitational interactions just as is the case for electromagnetic forces. This might appear to be irrelevant on pragmatic grounds as the planets move around the Sun at less than 1/1000 the speed of light, and relativistic effects are nugatory at such 'low' speeds. Nonetheless there was an issue of principle here, which Einstein resolved with his general theory of relativity, published in 1916.

THE COST OF FREE SPACE

Curved Space Time

Einstein had created his theory of special relativity by means of thought experiments involving electromagnetic radiation, light. Following this he did the same for gravity, which led him to general relativity.

Einstein had come to his original theory of special relativity under the assumption that there is no absolute state of rest. His general theory came from the thought that there is no absolute measure of force and acceleration. First, consider the problem for relativity when gravity is involved. Special relativity was based on the axiom that the velocity of light is a universal constant. Light has energy and as gravity acts not just on mass but also on energy in all its forms, then a light beam should be deflected by gravity as it passes near to a large object such as the Sun. As gravity fills the universe and light beams are being disturbed always, the principles of relativity, which assumed that light travels in straight lines at constant velocity, seemed only to be able to survive if gravity could somehow be turned off. Einstein realized this was a problem very quickly and it would take him ten years to solve it completely. His great insight came when he realized that gravity *is* effectively switched off for the case of a body that is in free fall. This means that there is no net force acting on the object and hence it moves at a constant velocity.

A lump of falling rock has no weight. If you catch it, what you perceive as weight is the force that you have to apply to stop it falling to ground. It is the solid floor that stops us falling to the centre of the Earth. It is the resistance of the floor, the force that it exerts to prevent our fall, that we feel as our own weight. Were the floor and the earth just vapour, we would fall to the centre of the planet, weightless.

This was the starting point of another of Einstein's thought experiments.

Suppose that you are inside a cabin that is in free fall, and there are no windows to look out of. This could be a broken elevator or, less traumatic, a satellite orbiting the Earth. In the latter case the satellite and you are in free fall but also travelling 'horizontally' at such a speed that the Earth's curvature makes the ground fall away from you at the same rate as you are falling towards it. In either example, in your immediate surroundings there would be no sense of any gravitational force. For example, if you let go of a ball, it will be pulled towards the Earth by gravity at exactly the same rate as you are, and so will remain stationary relative to you. Astronauts appear to float in their cabin for the same reason; they and the satellite are 'falling' at the same rate. Although we happen to know that the astronauts are falling in the gravitational field of the Earth, they feel no force and within their enclosed surroundings have every right to regard themselves as being at rest. Einstein had realized that in effect gravity has vanished in a free-falling 'weightless' state.

This also applies to light beams. A light beam will be attracted towards a massive body at the same rate as are conventional massive things. This was verified during the total eclipse of the Sun in 1919, when distant stars were seen to be displaced from their 'normal' position due to their light being deflected as it passed through the Sun's gravitational field.* Were you, trapped in the free-falling cabin, to shine a torch horizontally relative to its floor, a precision measurement made by someone on the ground would detect that the light beam had bent subtly as it 'falls' under gravity. In the brief moment that it crossed from one side of the cabin to the other, it will have fallen towards the Earth by the same amount as the opposite wall has fallen. Consequently, within the cabin it would appear to have travelled in a straight line; once again, everything is in accord with your interpretation that you are at rest in a force-free environment.

Suppose that you are in one of a convoy of spacecraft, each carefully positioned a kilometre apart and falling towards the Earth. Although astronauts in such a case could regard themselves as being at rest or in uniform motion in parallel straight lines, after a while they would begin to notice that all the craft were getting nearer to one another. The reason for this is that each is in free fall towards the centre of the distant planet, their trajectories converging towards that singular point. Einstein had

---

* Einstein had realized this as early as 1911 before completing his theory of general relativity. By 1916 he had the full theory, which revealed that the effect would be twice as large as he had originally supposed, as a result of both space and space-time being warped. An attempt to test his (erroneous) prediction during the total eclipse of 1915 had been abandoned due to the world war.

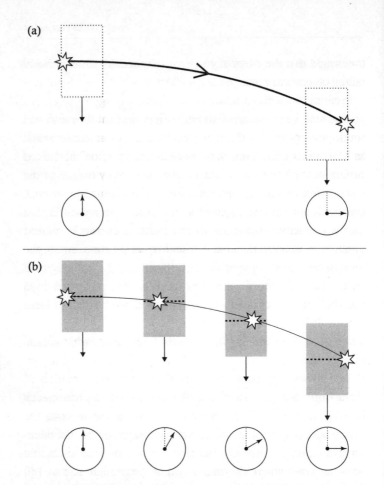

Figure 5. A flashlight halfway up the wall of a falling box is turned on. Its light beam crosses the box, the effects of gravity pulling both the beam and the box towards the ground. In (a) we see its curved path as viewed from the ground during the few nanoseconds that have elapsed. In (b) we see this same sequence as perceived by someone inside the box. As both the box and light beam fall at the same rate, the light beam appears to cross the box in a horizontal straight line.

the insight that the effect of gravity is to make the paths of freely falling objects converge.

Profound insights came when he saw an analogy between this picture and the convergence of lines of longitude at the north and south poles of the Earth. When mapped on a flat surface, such as Mercator's projection, these lines are parallel; on the curved surface of the Earth the 'straight lines' are initially parallel at the equator but as one heads northwards they gradually converge, eventually all coming together at the pole. The reason is that the two-dimensional surface of the Earth is curved in a third dimension. Einstein then made his remarkable extrapolation: the lines of free fall in a gravitational field are like lines of longitude on a 'surface' that curves in some higher dimension. He imagined that the three-dimensional 'surface' of space is stretched by large masses. It is free-fall motion along these curves that we perceive as deviating from 'straight' lines and interpret as the action of the force of gravity.

To see how Einstein incorporated this in his description of space-time let's go to a two-dimensional example. First recall Pythagoras' theorem that the square of the distance along the hypotenuse of a right-angled triangle is the sum of the individual squares along its perpendicular sides: $s^2 = x^2 + y^2$. This is true on a flat sheet where the angles within the triangle add to 180 degrees but is not in general true on a curved surface. This is most easily seen if we imagine taking a round trip, the first leg of which is along the equator from the Greenwich meridian to 90 degrees east. At this point turn left through 90 degrees and head north all the way to the north pole. If you now turn left through a further 90 degrees and head south (all directions are south from the north pole!) you will be travelling down the Greenwich meridian, eventually arriving at your original starting point on the equator

Figure 6. A view of the earth with a triangle superposed. One side is from equator to N. pole along Greenwich meridian; the base is along the equator running from Greenwich meridian to 90 degrees east (or west); the third side goes from equator to N. pole along the 90-degree east (or west) line of longitude.

to complete a triangle which contains three right angles. That the angles total more than 180 degrees is already an indication that you are not in flat space; it is obvious too that Pythagoras' theorem does not apply either—which of these three sides is the hypotenuse?!

There are other surprises when you live on a curved surface: what is a straight line when all lines must curve in at least one dimension?

The shortest distance between two points on a flat surface is a straight line. Einstein realized that it is the concept of shortest distance that is fundamental; in space-time that is curved by gravity, light follows the shortest path between any two points. On the Earth's surface these shortest paths are known as great circles. To fly from London at 55 degrees north to Los Angeles at nearer 30, you might naively expect to head in a south-westerly direction, whereas your flight by a great circle will go north-west towards and over Greenland. These great circles are more formally known as geodesics, meaning 'earth dividers'. The formula relating distances around triangles is more complicated than the Pythagorean form and requires knowledge of how the surface curves, how the 'metre' lengths relate to angles, or in the jargon, knowledge of the 'metric'. Einstein's goal of building a theory of gravity as curved space-time required answers to two questions. (i) Given some arrangement of matter, what is the metric of space-time that arises? (ii) Given the form of the metric, how do bodies move around?

If there was no matter present, the metric gives the relation that we saw already, $s^2 = x^2 + y^2 + z^2 - c^2 t^2$, and space-time is said to be flat. When matter is present the relationship between distances and time is changed and space-time is curved.

Evidence for this warping of space-time in our solar system most famously came from the orbit of Mercury which, like all planets, is an ellipse, but whose perihelion noticeably precesses. Being the nearest to the Sun it feels the strongest gravitational force, moves the fastest, and is most susceptible to the effects of relativity. The curvature of space makes the distance around the

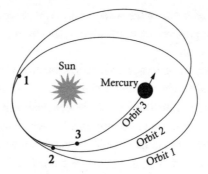

Figure 7. The advance of the perihelion of mercury. The marks 1, 2, 3 denote the points of closest approach on successive orbits.

Sun slightly different from its flat-space Newtonian value, which has the consequence that after completing a circuit, the trajectory does not end up in quite the same place as it would in Newton's picture. The result is that Mercury's orbit differs from year to year, in agreement with Einstein's theory.

For Einstein, space-time appears like an elastic solid, such as a rubber sheet. The force of gravity in this picture arises when a large mass, such as the Earth or Sun, is present at rest in the medium and distorts it. If the mass accelerates, for example two stars orbiting around one another, or when a star suddenly collapses and explodes as in a supernova, the theory implies that gravitational waves will spread out in the medium as when an earthquake spreads out seismic waves in the solid earth.

This prediction that gravitational radiation can occur still remains to be verified experimentally, in the sense of waves being detected, but there is indirect evidence for it. Two stars, known as the binary pulsar PSR 1913+16, orbit around one another every 7 hours and 45 minutes. The pulsar emits electromagnetic radiation in pulses, like a lighthouse beacon, every six-hundredths of a second. A lighthouse flash is what you observe because you see the circling beam only when it points at you, and you see nothing when it is oriented elsewhere. The six-hundredths of a second interval between successive flashes implies that the pulsar rotates seventeen times each second. In Einstein's theory such a system will emit energy in the form of gravitational waves and the time it takes to orbit will slowly fall. Such a change was measured by the astronomers Joseph Taylor and Russell Hulse and found to agree with what Einstein would expect; for this they won the Nobel Prize in 1975.

With this confirmation of Einstein's theory, we are led to a picture of space-time as acting like an elastic medium, which is reminiscent of the very ether that Einstein's work on electromagnetic radiation, his special theory of relativity, had done so much to eradicate. However, relativity does not imply that there is no ether, only that any stuff in that ether must behave in accord with the principles of relativity! An example of 'ether' is an electric field, which you cannot see unless you make it oscillate: then you literally can. A relativistic ether requires both electric and magnetic fields, changes in which travel at light speed. Analogously for the ether of a gravitational field, gravitational waves—ripples in the metric of space-time—also travel at the universal speed of light.

Gravity and Curvature

'Flat' space is space in which parallel lines never meet, which is all that Euclid and Isaac Newton knew; in curved space such lines are focused towards one another, the rate that the geodesics converge being a measure of the amount of curvature. Einstein formulated his theory of general relativity by relating the curvature to the gravitational field. This is what he did.

Space and time, or electric and magnetic fields, are only cleanly separated with respect to one observer; for another in relative motion they are entwined such that in space-time, electromagnetism is the only true invariant. Similar remarks apply to energy and momentum: it is 'energy-momentum' that acts as the relativistic measure of motion in space-time. Einstein had shown this in his 1905 theory of special relativity, which took no account of gravity. He had also shown, with $E = mc^2$, that mass is a form of energy; in his relativistic theory of gravity he generalized Newton's theory, where mass is the source of the force, by writing equations that related the density of energy-momentum to the curvature of space-time.

Note that I used equations in the plural. Curvature shows how a line deviates from one direction to another in any of the four dimensions of space-time, and to keep track of this required a separate equation for each possible combination of starting and finishing coordinate.

The amount of curvature is proportional to the energy-momentum density, and to the intrinsic strength of the gravitational force as defined by Isaac Newton 300 years ago, and inversely proportional to the fourth power of the velocity of light $(1/c^4)$. This makes sense: if gravity were stronger (weaker) the curvature caused by a given amount of energy-momentum would be more (less); and if c had been infinite, as Newton thought, then $1/c^4$ would have been zero and the amount of curvature would vanish, which is the same as saying that space-time is flat. This agrees with Newton's picture of space-time, an arena in which bodies move without affecting space or time and where parallel lines never meet. For Einstein by contrast mass and energy determine the shape of space-time. So Einstein's theory includes Newton's theory of gravity as a special case: it corresponds to c being infinite. For Einstein, signals travel no faster than c and there is no such thing as simultaneity whereas for Newton gravitation acted instantaneously, as if c were infinitely large.

In flat space-time, light beams follow straight lines, which is another way of saying that they follow the shortest path. In general relativity, light beams still take the shortest path. This is a familiar property in optics involving different media. It is the shortest 'optical path', in effect the minimum number of oscillations or shortest time, that causes the bending that we call refraction, as when a stick placed in water at any angle other than perpendicular to the surface will appear to be bent. It is the source of the rainbow, as light splits into different colours when it meets a surface of air and water or glass, as in a prism. This occurs as a result of the different colours, which correspond to different frequencies or rates of oscillation, each seeking their independent shortest optical path. So it is in space-time for bodies: a comet deflected by the Sun is following the path that minimizes the time it takes to pass from deep space on one side of the solar system to far away on the other.

An observer on Earth interprets the curved path of the comet as due to the gravitational force exerted by the Sun. Were Einstein on the comet, he would insist it was in free fall, effectively at rest and free of forces. It is thus travelling in a path that in flat space-time would be a straight line, in accord also with Newton, but in curved space-time is a curve.

In principle one could measure the curvature of space-time by forming a triangle with three light beams. Do the angles of the triangle add up to 180 degrees or do they exceed or fall below this value that we are used to in flat space? In the simple example that motivated all this, of falling towards the centre of the Earth, two light rays would converge together like lines of longitude and a triangle of beams would exceed 180 degrees; space would be revealed to be curved, but 'curved in what?' Recall that Einstein's original inspiration came from the two-dimensional surface of the Earth, which is curved in a third dimension; the converging paths of the spacecraft or light beams are curved in a higher dimension, at least mathematically. Loosely speaking, three-dimensional space is curved in the fourth dimension of time.* We can begin to picture this if we start with the simpler case of light travelling through flat space-time where there is no gravity.

An essential foundation of relativity theory is that light travels at constant speed for all. Nonetheless, when I speed towards or away from a light source, there is something that changes: as the pitch of a car horn rises or falls as the car rushes towards or away from you, so does the colour (the frequency or 'pitch') of light alter, being red shifted when the source is rushing away and blue shifted if heading towards you, a phenomenon known as the Doppler Effect. What we perceive as colour is a result of the different frequencies with which the electromagnetic fields can oscillate back and forth, and frequency is a measure of the beat of time. When light passes through gravitational fields, there is a further effect, and it is this that is at the source of the curvature of space in Einstein's picture.

When a ray of light passes through the gravitational field of the Sun, I will see its path curving. Light beams that are falling

* This is actually oversimplified as space and time are relativistically intertwined into space-time. Visualizing the full message of the mathematics is mind boggling. However, if we choose the perspective of one observer we may at least begin to understand some of what is happening with the concept of 'curved in time'.

towards the source of a gravitational force, such as the Sun, a neutron star, or a black hole, are converging towards one another, like the spacecraft that we met earlier. According to general relativity, not only does their motion shift their colour as perceived by a stationary observer, but gravitational forces do also, the frequency of oscillations of the electromagnetic fields becoming increasingly shifted towards the red in ever stronger gravitational fields. As the beams approach the source of the gravitational attraction, distant watchers will perceive them to be red shifted more and more. The frequency of the oscillations, their natural time clock, slows. Were the light beam to approach the edge of a black hole, the frequency would slow to nothing; in a sense, time would stand still such that from the perspective of an observer on Earth, it would take infinite time for the beam to enter the hole as it became redder and fainter. For the light beam itself, nothing appears to be happening as it is in free fall. Other light beams converge ever closer upon it, indeed within the black hole all trajectories curve so tightly that those heading outwards never cross the boundary; light never escapes to the outside and the hole appears black. Under the influence of gravity, light beams are travelling along geodesics in a universe where time is being increasingly stretched. It is this distortion of the time dimension that leads to the appearance of curvature for the paths in the other three dimensions of space. If you can extend this simple picture of the stretching of time, and imagine it intertwining with that of space in a relativistically invariant space-time, then you have a better imagination than I; suffice to say the mathematics of Einstein's equations is keeping the accounting correct while the underlying physics is the time stretching that occurs if gravity is 'switched off' in free fall.

Expanding Universe

Although the basic idea is intuitively simple to visualize, solving Einstein's equations is not, and even today nearly a century after first being written down they have only been solved in a limited number of cases. The simplest is when there is no energy-momentum, in which case there is no curvature: the universe is flat. There are also solutions in which space-time contains no matter and yet is not flat. Whereas this runs counter to the naive expectations in the philosophies of previous centuries, in general relativity this can occur due to the fact that signals propagate at the finite speed of light, c, rather than instantaneously. If something happens that causes the distribution of energy suddenly to change, such as a supernova explosion or a star collapsing to form a black hole, the gravitational waves will radiate outwards at the speed of light. Gravitational fields are themselves full of energy and a localized ripple will cause further gravitational effects, waves of energy that spread onwards. If the original material source of the gravitational wave is removed, the wave can continue to spread. So one could imagine a region of the universe devoid of matter but whose space-time is rippling with gravitational waves. So much for the emptiness of the Void!

'Ripples in space-time' begs the question of what this means in any absolute sense and how they can be detected. As an earthquake gives waves on the ground, disturbing the geodesics of Earth, so will gravitational waves cause oscillations in the geodesics of any photon beams, and in the space between atoms of any material bodies. Their effects are like tidal forces, stretching and pushing any existing matter into new shapes. While only indirect hints have been found so far (as in the binary pulsar example mentioned earlier), finding direct evidence for gravitational waves is actively on the scientific agenda. Detectors at laboratories thousands of kilometres apart are being linked electronically to make a coherent large-scale experiment code-named LIGO for 'laser interferometer gravity-wave observatory'. Detectors on widely separated satellites also are being planned with the acronym LISA for 'laser interferometer space antenna'. When a gravitational wave hits a bar that is over a kilometre in length, the bar will shrink slightly, perhaps by less even than the size of a single atom. By reflecting laser beams from mirrors, changes in distance of atomic scales can be revealed. Gravitational waves are expected from colliding stars, black holes, supernovae, and other catastrophic events, and it is hoped to be able not just to detect the waves but also to identify the nature of their sources. Scientists are even hopeful of detecting faint echoes of the Big Bang.

Having written the equations, Einstein wanted to see what they implied for the universe, and to do so he assumed that the universe is uniform in all directions. This led to a startling conclusion: the space-time grid of the universe cannot remain uniform, static; it has to be changing. In effect the equations revealed that the gravitational attraction of all pieces of matter to all the others throughout an infinite cosmos is unstable, the slightest deviation from homogeneity leading to collapse. Two possible resolutions of this conflict occurred to him. One was that the universe is expanding—a solution that the equations allowed; however in 1915 the received wisdom was that the universe is static, unchanging, and so Einstein seized on another possibility. His equations allowed that in addition to the well-known inverse square law of attraction, the force of gravity could contain an extra component whose strength grows with increasing distance and acts as a kind of anti-gravity. Such an effect would be negligible over the size of the solar system or even our galaxy, but on the immense distance scales of the universe it could be significant and stabilize the cosmos. He called this the Lambda force, denoted by the Greek symbol Λ, also known as the Cosmological Constant.

What happened in the subsequent years is ironic. First, it turned out that the presence of Λ does not solve the problem; Λ does not make the universe static. Einstein described this as the greatest blunder of his life. It was a technical blunder and also a failure of intuition as within a few years Edwin Hubble's astronomical observations revealed that the universe contains galaxies which on the average are moving apart from one another. The further away they are from us, the faster they are receding, which is consistent with the picture that the universe is expanding. Such a behaviour is what Einstein's equations had actually predicted before he tried to prevent it by introducing the Λ force. To complete the irony, recent observations suggest that the expansion is gradually getting faster, as if there is some cosmic repulsive force at work. It is possible that it is the first proof that there is indeed a small Λ term.

It is as if all of space is filled with a strange sort of anti-gravity, which has become known as dark energy. Its effects were masked in the early, small, compact universe but, as the universe expanded, the gravitational forces between its ever more distant galaxies were weakened to the point where the effects of the universal Λ energy began to win. That tip-over seems to have occurred about 5 billion years ago.

The accelerating expansion rate of the universe that has been observed suggests that Λ is very small, indeed incredibly small; compared to Newton's measure of the gravitational force, it is some 10^{126} times smaller.* If it had been large, theorists would be relatively comfortable; if it were not there at all, were zero, that too would fit with our understanding. The fact that every cubic metre of space is filled with dark energy in amounts that are incredibly tiny, and yet not quite zero, is a profound puzzle about the nature of the vacuum, the 'cost' of free space.

* To give an idea of the size of 1 followed by 126 zeroes, this number exceeds by factors of trillions the number of protons in the entire observable universe.

7

THE INFINITE SEA

The Quantum World

In 1687 Isaac Newton laid down the first universal laws of gravitation in his *Principia*. By the mid-nineteenth century, James Clerk Maxwell had united a multitude of electric and magnetic phenomena with his elegant theory of electromagnetism. 'There is nothing new to be discovered in physics now,' William Thomson, Lord Kelvin, asserted at the British Association meeting in 1900. Within five years Einstein had invented relativity theory. Ironic then that Albert Michelson, whose own experiments had helped form the paradoxes that led to that new world view, had also insisted that 'The grand underlying principles have been firmly established; further truths of physics are to be looked for in the sixth place of decimals.'[1]

Nature repeatedly reveals the limits of our collective imagination. The discoveries of relativity, the nuclear atom, and the rise of quantum mechanics showed how naive Lord Kelvin and Michelson had been. Newton and Einstein's mechanics are without peer in describing the behaviour of large bodies, from entire galaxies to falling apples, and even of beams of light. The former are nearer to direct experience; that of light was less intuitively obvious. The discovery that on the atomic scale we need to use quantum mechanics, and that this seems to reveal a will o' the wisp world of uncertainty, has become the other great foundation of modern science and one that is far from intuitive. It will turn out to have profound consequences for our attempts to understand the Void. Indeed, quantum mechanics seems to imply that Aristotle may have been correct; far from a vacuum being empty, it is always seething with activity. So first, let's meet the ideas of the quantum and try to understand how they relate to the ideas of Newton and Einstein.

Human beings are huge compared to atoms. Our senses have developed such that they make us aware of the macroscopic world around us. Our ancestors' eyes developed so they are sensitive to the optical spectrum; they needed to see potential predators and had no need to view radio stars or atoms. To see atoms required special microscopes that were only developed within the last hundred years and began to reveal phenomena that ran counter to the known laws of physics. For example, whereas billiard balls bounce off one another in a determined way, beams of atoms will scatter in some directions more than others, forming areas of intensity or scarcity like the peaks and troughs of water waves that have diffracted through an opening. It is the macroscopic world that as young children we are first aware of, and around which we build our intuition. Our subsequent expectation of how things should behave is based upon that; wave-like atoms are not part of the normal scenery.

Nothing of atoms was known in the seventeenth century when Isaac Newton encoded the mechanics of macroscopic bodies, which were later refined by Einstein and which have formed the axioms of our story so far. However, this view of nature is a gross one. For objects that consist of vast numbers of atomic particles the mechanics of Newton and of Einstein are adequate but not fundamental. The individual particles obey more basic rules that are often strange to our senses, 'strange' because, for instance, one cannot know both the precise position and motion of individual atoms. If individual atoms had awareness, their intuition would have developed from such experiences; this would be the nature that they know and it would appear—natural. However, self-awareness involves vast numbers of atoms. When large numbers of atoms become organized, simple regularities can emerge, giving properties to the organized collection that individual atoms or small numbers of them do not have. Human consciousness may be one example; others include the magnetism of metals and the property of superconductivity that emerge for macroscopic collections of atoms but which individual atoms do not have, and the phases of solid, liquid, and gas as in ice, water, and steam that arise from the different ways that the same atoms or molecules manage to organize themselves (we shall develop such ideas further in Chapter 8 when we meet the idea of phase transitions and consider whether there is a unique vacuum). In such situations, from the underlying fundamental behaviour does a hierarchy of physical laws emerge.*

* See R. B. Laughlin, A *Different Universe*, (Basic Books, 2005) for an extensive description of emergent phenomena throughout physics. He gives particular emphasis to the idea that Newton's laws are descriptive and not fundamental, and that our difficulties with quantum phenomena are due to trying to interpret them in terms of Newton, whereas we should be accepting Newton as emergent from the quantum.

The reason that predictive science is possible, even though the fundamental equations may be unknown or, if known, be impossible to solve, is because it is not just atoms and molecules that respect organization: laws that operate at the level of individual atoms become organized into new laws as one moves up to complex systems. The basic equations that control the individual atoms are known, but solving them is possible in only a few simple cases and deriving the existence of solids and liquids all but impossible. Yet this does not prevent engineers from designing solid structures or hydraulic systems. The laws of electrical charges beget those of thermodynamics and chemistry; in turn these lead to the laws of rigidity and then of engineering. The derivation of the liquid state for this or that substance from first principles may be lacking, but there are still general properties that liquids have that transcend these. Liquids will not tolerate pressure differences from one point to another other than those due to gravity; this is the principle behind the mercury barometer and all hydraulic machinery. This is a property of the organized liquid state and the detailed underlying laws at the atomic level are essentially irrelevant. It begat the phenomena that led Galileo and Toricelli to their discoveries of liquids and vacua with which we began our journey.

It is this hierarchy of structures and laws that enables us to understand and describe the world; the outer layers rely on the inner yet they each have an identity and can often be treated in isolation. Thus can the engineer design a bridge without need of the atomic physics that underpins the laws of stress and strain.

Newton's laws of motion—things move at a constant speed unless forced to alter their motion; the same amount of force accelerates a heavy thing less than a light one; acceleration takes place in the same direction as the force that causes it—underpin all of engineering and technology. In 300 years of careful experimentation their only failures are when applied to objects moving near to the speed of light, whence they are subsumed in Einstein's relativity theory, and at atomic length scales, where the laws of quantum mechanics replace them.

Our immediate experiences are of bulk matter and our senses are blind to the existence of atoms, but clues to the restless agitation of the atomic architecture are all around. As I watch my plants grow I don't see the carbon and oxygen atoms pulled from the air and transformed into the leaves; my breakfast cereal mysteriously turns into me because the molecules are being rearranged. In all cases the atoms are calling the tune and we lumbering macro-beings see only the large end-products. Newton's laws apply only to the behaviour of those bulky things.

Two hundred years after Newton, experimental techniques had progressed to the point that the atomic architecture was beginning to be recognized. By the start of the twentieth century numerous strange empirical facts about atomic particles began to accumulate that seemed incompatible with Newton's clock-work, such as the wave-like behaviour of atoms mentioned earlier. If we try to describe this weirdness using familiar Newtonian language, we fail.

The solution to the conundrum is 'a beautiful case history of how science advances by making theories conform to facts rather than the other way round'.[2] The laws of quantum mechanics, which are the mechanics of very small things, were discovered in the 1920s. Quantum mechanics works: it makes predictions that in some cases have been confirmed to accuracies of parts per billion. Yet it creates mind-bending paradoxes that some charlatans exploit to convince the public that scientists seriously consider parallel universes where Elvis lives, or that telepathic communication is possible.

One of the apparent paradoxes that concerns us is that after removing matter, fields, everything to reach a void, the emptiness that ensues at large scales is also a collective effect. When viewed at atomic scales, the Void is seething with activity, energy, and particles.

Waves and Quantum Uncertainty

All of quantum mechanics derives from one fundamental property of nature: it is not possible to measure both the position and momentum of a particle with arbitrary precision. If you know the position perfectly, then you know nothing at all about its momentum, and vice versa. In general there is a compromise. If the position of a particle is known to be within some distance r of a point, then its momentum must be indeterminate by at least an amount p where

$$p \times r \sim \hbar$$

and $\hbar$ is a constant of nature known as 'Planck's constant' (actually divided by 2π). Its magnitude* is so small as to be irrelevant for macroscopic objects, but for atoms and their constituents it is what controls their behaviour.

A similar uncertainty applies to time and energy (I said 'one' fundamental property above because in space-time the quantum see-saw between space and momentum is matched by time and energy). This implies that energy conservation can be 'violated' over very short time scales. I put 'violated' in quotes because one

---

* $\hbar$ is pronounced 'h-bar', its magnitude is $\hbar = 1.05 \times 10^{-34}$ Js $= 6.6 \times 10^{-22}$ MeVs.

cannot detect it; this is the nub of the inability to determine energy *precisely* at a given time. Particles can radiate energy (e.g. in the form of photons) in apparent violation of energy conservation, so long as that energy is reabsorbed by other particles within a short space of time. The more that the energy account is overdrawn, the sooner it must be repaid: the more you overdraw on your bank account, the sooner the bank is likely to notice; but pay it back before being found out and everyone is satisfied. This 'virtual' violation of energy conservation plays an important role in the transmission of forces between particles. In the quantum picture of the electromagnetic field, it is virtual photons, quantum bundles or 'particles' of light, that flit across space-time and transmit the forces between remote objects.

Notice how I slipped in 'photons' as 'particles' of light here. Is light not a wave? The dual nature of wave or particle goes back to Isaac Newton. Light rays act as if composed of streams of particles: travelling in straight lines, leaving sharp shadows, deviating at the junctions of different media, as between air and glass, according to the classical rules of geometrical optics. Yet light also shows a distinct wave-like character: the edges of shadows are not so sharp; when scattered through pinholes dark and light bands known as interference fringes can arise. The fact that two overlapping pieces of light can under certain circumstances cancel out and give darkness, as in such displays, is most readily understood in terms of two waves meeting; when two peaks coincide there is a big peak, or intense brightness, but when a peak and a trough meet, they cancel, giving darkness.

In 1900 Max Planck had shown that light is emitted in distinct microscopic 'packets' or 'quanta' of energy known as photons, and in 1905 Einstein showed that light remains in these packets as it travels across space. It was in his theory of energy quanta that Planck had introduced his eponymous Planck's constant traditionally abbreviated to the symbol h (the combination $h/2\pi$ being denoted $\hbar$). This was the beginning of quantum theory and its immediate success was in explaining how atoms could survive.

The electron in a hydrogen atom is apparently orbiting around the central proton at a speed of $1/137$ of that of light. An orbit of 10^{-9} metres at a speed of about a thousand kilometres per second implies some million billion circuits each second. According to Maxwell's theory such an electron should emit electromagnetic radiation so readily that the moment such an atom formed, the electron would immediately spiral into the nucleus in a blaze of light. So how do atoms exist? The discovery that radiant energy is quantized led Neils Bohr to propose that the energies of electrons in atoms are also quantized: they can have only certain prescribed energies. Restricted to these particular energy states, electrons do not radiate energy continuously and so do not smoothly spiral inwards. Instead, they can only jump from one energy state to another and emit or absorb energy to keep the total amount of energy constant (over long time scales energy is conserved). Once in the lowest energy state of all, they have nowhere lower to go and so they remain there making a stable atom. You may already be suspicious that this is a solution by dictat: it is stable because it is stable. However, if we adopt the wave picture it is possible to imagine why.

Bohr proposed that Planck's constant h controlled the permissible energies of electron orbits in atoms. In the modern picture, not just light but also an electron has a wave-like character, and its wavelength and momentum are linked by the same quantity h. Now apply this idea to the simplest atom, hydrogen, whose nucleus is encircled by just one electron. The electron waves cancel out and are destroyed in any paths where they do not

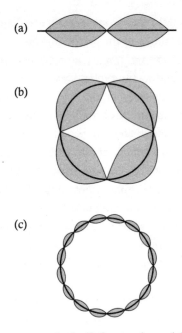

Figure 8. Electron waves in the Bohr atomic model.

'fit'. This is illustrated in Fig. 8. In Fig. (a) an electron moving along the path—is represented by a wave. Now imagine a complete wavelength bent into a circle. When the wave fits the circle precisely, this is the first allowed orbit; if the waves do not fit like this, they die out. Two wavelengths completing the circle as in Fig. (b) gives the second Bohr orbit, which has higher energy than the first and higher energy orbits correspond to larger numbers of wavelengths fitted into the circumference (Fig. c). Remarkably this simple picture fits with what we know of atoms.

No energy is radiated when the electron stays in an orbit but energy is emitted if it jumps from a high-energy to a lower-energy state. By assuming that this radiated energy was converted to light, Bohr calculated the corresponding wavelengths and found that they matched precisely the mysterious spectrum of hydrogen. Planck's quantum theory, applied successfully to radiation when Einstein postulated the photon, had now been applied to matter with equal success by Bohr.

An essential feature of this is that the quantum theory is taken to imply that wave-particle duality is a property of all matter: the electron, which we think of as a particle, is really a quantum bundle of an ' electron-field' which acts with wave-like properties. Weird as it may sound, that is how it is; electron microscopes exploit this wave-like character of electrons.

What are these waves and how do they relate to the uncertainty principle that we met above? Such questions have plagued science ever since the birth of quantum theory. Einstein and Bohr, among others, argued at length about the meaning of quantum theory, so forgive me if I do not profess to have the answers. Here is how I try to come to terms with it; if you prefer some other then please proceed with that as there is no agreed wisdom as to any 'official' explanation.

At the purest level one just has to accept the uncertainty principle and its implications. However, it is always more comforting when we can form a mental model with properties that the theory has, as then we can develop intuition about its behaviour and implications. The position and momentum uncertainty does have an analogue that we are familiar with. Draw lots of dots to form a wave with a fixed wavelength; then if we identify position as the location of a given dot in the wave, and momentum as the wavelength; this is an analogue of the uncertainty principle at work. According to quantum mechanics, the higher the momentum so the shorter is the wavelength. Suppose I know the position precisely; then all I have is a single dot and it is impossible to know what the wavelength will be; it could be anything you want. If I have a few dots forming the beginning of the wave, then I will begin to see if the wavelength is small or large, and only after I have a complete wavelength will I be able to say with absolute certainty what its value is. However, the price of this certainty in knowing the wavelength is giving up knowledge of position to any better precision than the length of the wave.*

One sees here that it is an oxymoron to define the position of a wave; it only becomes a known wave when one measures over its full wavelength. If this at least opens your mind to accept that there are familiar concepts for which position and another quality cannot both be meaningfully defined with precision, then one is beginning to appreciate the nature of the quantum world. The fact that waves have these properties makes them very useful as mental models of what is happening. However, in my opinion that is all that they are: mental models.

A Seething Vacuum

Imagine a region of vacuum, for example a cubic metre of outer space with all of the hydrogen and other particles removed. Can it really be devoid of matter and energy? In the quantum universe the answer is no.

Having the precise information that there is no particle at each and every point implies knowing nothing about motion and hence of energy. You may remove all matter and mass, but quantum

---

* Mathematically this is realized by Fourier analysis—the representation of any curve, or even an abrupt spike, as a superposition of waves with different wavelengths. A singular spike at a precise location is equivalent to a sum over an infinite set of waves of all wavelengths.

uncertainty says there exists energy: energy cannot also be zero. To assert that there is a void, containing nothing of these, violates the uncertainty principle. There is a minimum amount known as zero point energy, but that is the best you can do. It is possible to visualize this by considering a pendulum consisting of just a few atoms.

The precise speed of a particle can only be determined if its position is unknown. This implies that a small cluster of molecules suspended by a thread of atoms and swinging like a pendulum could never come completely to rest, hanging vertically, with the ball of molecules stationary at the lowest height, or 'zero point'. Instead, quantum uncertainty implies that it must wobble slightly around this position. This phenomenon is called zero point motion.

As it swings under the influence of gravity, the higher above the zero point the molecules are, so the greater is their potential energy. At the top of the swing the potential energy of a macroscopic pendulum is at its maximum, the kinetic energy being zero; conversely, at its lowest point the potential energy is zero and the kinetic energy is maximal. Things are more subtle for a 'nanoscopic' quantum pendulum. If we minimize the potential energy by restricting the pendulum's ball to be at height zero, its state of motion and hence kinetic energy become indeterminate. Conversely, minimize the kinetic energy by having the pendulum at rest, and its height above zero becomes unknown. Quantum mechanics implies that there is a minimum sum of kinetic and potential energies that can be achieved: both cannot simultaneously be zero. This minimum amount is the zero point energy of the atomic assembly.

For a macroscopic pendulum, such as in an antique clock, this zero point energy is too small to notice. However, for clusters of

Classical pendulum

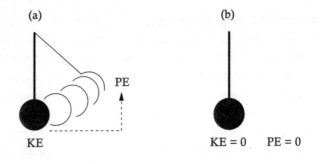

Quantum pendulum

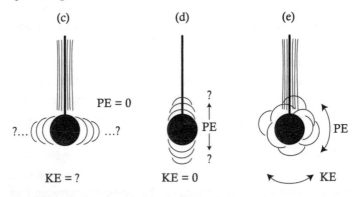

Figure 9. (a) The pendulum starts high up at rest: its potential energy (PE) is big and its Kinetic energy (KE) is zero. Gravitational force swings it downwards; at the lowest point where it has no PE, it will have its maximum KE. Throughout the swing the sum of PE + KE is a constant. (b) It is possible to hang the pendulum vertically and at rest. The PE is zero as is the KE. The total energy is therefore zero. (c) For a quantum pendulum we cannot have PE and KE simultaneously zero. Hanging at the lowest point with PE = 0, the motion is indeterminate, and so the KE is unknowable. This is 'zero point motion'. (d) Alternatively, if the pendulum is at rest with KE = 0 its position and hence PE are undetermined. (e) There is a minimum sum possible for PE + KE known as the zero point energy.

a few atoms and molecules this minimum energy is comparable to the total energies of these groups of particles themselves. The zero point energy is then manifested by motion, for example of the atoms within molecules and of the individual molecules within the bulk cluster. Thus while the motion of molecules in a substance gives rise to what we call temperature, the higher the temperature so the more agitated their motions, the quantum theory implies that there will remain an intrinsic zero point energy even as one approaches the absolute zero of temperature.* One implication is that it is impossible to achieve absolute zero of temperature where everything is both frozen in position and without momentum and energy.

The remarkable thing is that this applies to a finite size of space, even if there is no matter in it. The consequence is that a finite region of empty space, 'empty' in the sense of having all matter removed, will be filled with energy. All finite volumes of whatever size are subject to fluctuations in energy. For macroscopic volumes the effect is too small to notice, but for very small volumes the energy fluctuations are big.

As two pieces of light can cancel to zero due to their wave-like character, so can zero turn to two counterbalancing somethings. The Void may have no electromagnetic fields on the average, but fluctuations driven by the zero point phenomenon are always present with the result that there is no such thing as literally empty space. In the modern perspective, the vacuum is the state where the amount of energy is the minimum possible; it is the state from which no more energy can be removed. In scientific jargon this state of vacuum is called the 'ground state'. Latent within the laws of nature are excited states, with energy densities corresponding

* This is −273 degrees Celsius, which is 0 K, zero degrees Kelvin.

to one, two, or even billions of material particles or radiation. You can remove all of these real particles until you reach the ground state, but the quantum fluctuations will still survive. The quantum vacuum is like a medium, and from what we know about the ground states in macroscopic collective systems, further surprises can be expected for the properties of the quantum vacuum, as we shall see in Chapter 8.

First we need to be convinced that zero point energy is real and not some artefact of mathematics. A physical consequence was suggested in 1948 by Hendrik Casimir and, after years of attempts, was finally demonstrated experimentally in 1996.

The Void is a quantum sea of zero point waves, with all possible wavelengths, from those that are smaller even than the atomic scale up to those whose size is truly cosmic. Now put two metal plates, slightly separated and parallel to one another, into the vacuum. A subtle but measurable attractive force starts to pull them towards one another. There is of course a mutual gravitational attraction of the one for the other, but that is trifling on the scale of the 'Casimir effect', which arises from the way that the plates have disturbed the waves filling the quantum vacuum.

The metals conduct electricity and this affects any electromagnetic waves in the zero point energy of the Void. Quantum theory implies that between the plates only waves that have an exact integer number of wavelengths can exist. Like a violin string vibrating between its fixed ends giving a tone and harmonics, only those waves that are in 'tune' with the gap between the plates can 'vibrate', whereas outside the plates all possible wavelengths can still exist. Consequently there are some waves 'missing' between the plates, which means that there is less pressure exerted on the inside of the plates than on their outward faces, leading to an overall force pressing inwards. Quantum mechanics predicts how large this force should be. Its magnitude is proportional to Planck's quantum, h (as it is a quantum effect), the velocity of electromagnetic waves, c, and inversely proportional to the distance d between the plates to the fourth power, d^4. This implies that the force vanishes as the plates become far apart, which makes sense as for infinite separation we are back with the infinite void for which there can be no effect. Conversely, the force will be larger when the two plates are very close; in such circumstances it is possible to measure it, verifying both its magnitude and variation with the distance of separation.

The force has been measured, the effect confirmed, and the concept of zero point energy in the Void established. The Casimir effect demonstrates that a *change* in the zero point energy is a real measurable quantity, even though the zero point energy itself is not available. The amount of zero point energy is actually infinite and some misinterpretations of the theory have led to suggestions in tracts such as *Infinite Energy* (*sic*) magazine that this is a source of power that has been overlooked by science until tapped by workers in cold fusion and the like. Zero point energy is not like this. It is the minimum energy that a system, or the vacuum, can have.

The zero point motion of electromagnetic fields is ever present in the vacuum. The zero point energy of the vacuum cannot be extracted or used as power; the vacuum is as low as it gets. Yet the effects of zero point motion can be felt by particles passing through the vacuum.

An electron in flight wobbles slightly as it feels the zero point motion of the vacuum electromagnetic fields. To reveal this we need some measurable reference and an electron trapped within a hydrogen atom is enough to show that the vacuum is far from empty. The electron in hydrogen is moving at a speed of about

1 per cent of the speed of light. The spectrum of hydrogen reveals the energy changes as electrons jump between different orbits in the atoms. The differences in energies between the various levels are manifested as the energy of the light that appears in the spectral lines.

Techniques that had been developed in radar during the Second World War enabled post-war physicists to measure the energies of the spectrum, and by inference of the electrons, to an accuracy of better than one part per million. This led to the discovery of the 'Lamb shift', named after Willis Lamb who first measured it in 1947; this subtle shift relative to what quantum mechanics expected if the vacuum were truly empty agrees perfectly with calculations that include the effects of fluctuations in an effervescing quantum vacuum.

While quantum mechanics makes precise statements about phenomena on subatomic length scales, it does so while ignoring the effects of gravity. No one has successfully combined the two great pillars of twentieth-century physics—quantum mechanics and general relativity—to make a mathematically consistent and experimentally tested unified theory. In practice scientists sidestep this as the two theories are each flawless in their respective arenas. Yet in the first 10^{-43} s of the Big Bang, the universe was so small and gravity so all embracing that a theory of quantum gravity would rule. Establishing what this is remains one of the major unsolved challenges in mathematical physics. However, we can appreciate the profound implications it will have for some of the problems that we need to answer. For example, our experience is that the dimensions of space and time are somehow different, at least in our ability to travel through them and to receive or process information. While this subtle difference is true as perceived by our macroscopic senses, and to our description of natural phenomena down to the scale of atoms and beyond, when in those first moments our universe was compressed into a distance scale of about 10^{-35} m, a quantum theory of gravity would intertwine space and time inextricably. In quantum gravity, space and time must somehow be 'the same'.

The complementary uncertainty between motion, momentum, and energy, and location in space and time, suggests that in quantum gravity there are fluctuations occurring in the fabric of space and time themselves. If we were to measure distances that are as small compared to a proton as that proton is to a human, or to record time scales as short as 10^{-43} seconds, we would find that Newton's matrix had evaporated into a space-time foam. I cannot imagine what this would be like, but science fiction writers love it.

There is general agreement that the quantum vacuum is where everything that we now know came from, even the matrix of space and time. As we shall see, the seething vacuum offers profound implications for comprehending the nature of Creation from the Void.

The Infinite Sea

The stability of matter and the periodic regularity in Mendeleev's table of the atomic elements are ultimately due to the fact that electrons obey a fundamental rule of quantum mechanics known as the exclusion principle: no two electrons in some collection can occupy the same quantum energy state. When Paul Dirac first realized that quantum theory implied that electrons can have positively charged 'anti'-electron counterparts known as positrons, he used this exclusion principle to make a model of the vacuum that would naturally give rise to such unusual entities. He proposed that we regard the vacuum as being far from empty: for Dirac it was filled with an infinite number of electrons whose individual energies occupy all values from negatively infinite up to some maximum value. Such a deep, calm sea is everywhere and unnoticeable so long as nothing disturbs it. We call this normal state the ground state, which is our base level relative to which all energies are defined: Dirac's 'sea level' defines the zero of energy.

Einstein's famous equation $E = mc^2$ can be rearranged to read $m = E/c^2$, which says that mass can be produced from energy. An electron and its antimatter twin, the positron, have the same mc^2 and equal but opposite signs of electric charge. So if the energy E exceeds $2mc^2$ it is possible for an electron and a positron to emerge. The energy fluctuations in the vacuum can spontaneously turn into electrons and positrons but constrained by the uncertainty principle to last only for a brief moment of less than $\hbar/2mc^2$, which amounts to a mere 10^{-21} s. This time is so small that light would have been able to travel only across about one thousandth the span of a hydrogen atom. Such 'virtual' particles cannot be seen any more than can the deviation from energy conservation that these fluctuations amount to. However, the implication that the vacuum is filled with virtual particles can be detected by careful and precise measurements.

An electrically charged particle, such as an electron or an ion, is surrounded by a virtual cloud of electrons and positrons.* One effect of these clouds is to modify the strength of the electrical forces between two charged objects. The finer the microscope with which we look, the more we become sensitive to the effects of

---

* It is also surrounded by all other varieties of charged particles and their antiparticles; the heavier they are, the more nugatory is their fluctuation, and so it is the electron and positron, being the lightest, that are the dominant players.

these virtual clouds in the vacuum. As an electron and positron pair fluctuate into and out of their virtual existence within only one thousandth part of an atomic radius, they can influence the force between the proton and remote electron in a hydrogen atom, which gives a small modification to the inverse square law of force, and also affect the magnetism of particles like the electron in calculable ways that agree with the data to a precision of better than one part in a hundred billion.

In Dirac's interpretation of the vacuum as an infinitely deep sea filled with electrons, if one electron in this sea were missing, it would leave a hole. The absence of a negatively charged electron with energy that is negative relative to sea level will appear as a positively charged particle with positive energy, namely with all the attributes of a positron. Fluctuations in the surface of the sea, in accord with the zero point energy phenomenon described earlier, could momentarily elevate an electron leaving a hole, appearing as a virtual electron–positron pair.

It is possible to make these virtual fluctuations visible by supplying energy to the atom. If a photon with energy greater than $2mc^2$ irradiates an atom, it is most likely that it will ionize that atom. However, it is possible that a virtual electron and positron are bubbling within the atom's electric field as the photon hits. In such a case the photon may eject them out of the atom, leaving the atom behind undisturbed. This phenomenon, known as 'pair creation', can be photographed in a bubble chamber leading to beautiful and enigmatic artwork as in Fig. 10.[3] The two virtual particles thus become real.

For Dirac, such antiparticles are holes left in the infinitely deep sea that is the vacuum. This picture also resolves what would otherwise be a paradox. If the vacuum were truly empty, then what would encode the laws of nature, the properties of matter,

Figure 10. Pair creation.

(a)

E = 0

E < 0

(b)

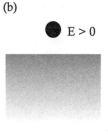

E > 0

(c)

(d)

E > 0

E = 0

E < 0

Figure 11. (a) The vacuum is filled with an infinitely deep sea of filled energy levels from negative infinity to some maximum. We define this configuration, the state of lowest energy, to have zero. (b) A state with positive energy, e.g. an electron with positive energy relative to the vacuum. (c) A hole in the vacuum. Absence of a state with negative energy and negative electric charge will appear as if a positive energy state with positive charge. This is Dirac's picture of the antiparticle of an electron: the positron. (d) A state with negative energy is empty and a positive energy state is filled. This could be a positive energy electron and the 'hole' is perceived as a positive energy positron. To produce this configuration energy must first be supplied to the vacuum. This energy could be donated by a photon whereby the photon has converted into an electron and a positron. A photograph of a real example of this process is seen in Fig. 10.

such that all electrons and positrons created 'out of the vacuum' have identical properties, with specific masses rather than emerging with a random continuum of possibilities? Protons and quarks and similar particles also satisfy the exclusion principle and fill an infinitely deep sea. It is the infinitely deep storehouse of the Dirac sea that provides us with the particles that we can materialize.

In this interpretation, the vacuum is a medium. It has profound connections with phenomena that occur in 'real' media, such as solids and liquids where vast numbers of atoms or particles organize themselves into different 'phases'. Thus the quantum vacuum is like the configuration with the lowest possible energy, the 'ground state', of a many-body system. We will see more of this in the next chapter. The implications are profound, including the possibility that the nature of the vacuum has not always been the same throughout the history of the universe. It also raises an interesting possibility: that one could *add* something to the vacuum and yet *lower* its energy. In such a case one would have created a new state of vacuum; the previous vacuum, which has higher energy than the true ground state, being known as a 'false vacuum'. The transition from the false to the new vacuum is known as a phase change. Theorists speculate, and experiments in high-energy physics may soon give the answer, that something like this happened early in the history of the universe at temperatures in excess of a million billion degrees (Chapter 8).

THE HIGGS VACUUM

Phases and Organization

In Chapter 6 we briefly met the idea of organization, in which large numbers of atoms and molecules can take on characteristics that individuals do not have. As the quantum void is filled with particles, it too can have unexpected properties that depend on how its constituents are organized. There are many familiar examples of organization, and as they have inspired modern ideas about the nature of the vacuum, I will start this chapter by describing some.

Emergence is said to occur when a physical phenomenon arises as a result of organization among any component pieces, whereas the same phenomenon does not occur for the individual pieces. Thus in art, the individual daubs of paint in an impressionist canvas by Monet or Renoir are randomly shaped and coloured, yet when viewed from a distance the whole becomes organized into a perfect image of a field of flowers. It is the very inadequacy of the individual brush strokes that shows the emergence of the painting to be a result of their organization. Analogously, individual atomic 'brush strokes' can form an organized whole capable of things that individual atoms, or even small groups of atoms, cannot do. Thus one proton or electron is identical to another and all they can do individually is to ensnare one another by their electrical attraction to make atoms; the electricity within atoms enables groups to join, making molecules; bring enough of them together and they can become self-aware—such as you reading this.

Certain metals can expel magnetic fields when cooled to ultralow temperatures, giving what is known as superconductivity, yet the individual atoms that make the metal cannot do this. The emergence of solids, liquids, and gases from a large collection of molecules, such as the H_2O of water, ice, and steam, is an everyday example. We take for granted that the solid floor of a plane flying at a height of 10,000 m will not suddenly lose its rigidity and release us into the clouds below. Eskimos likewise trust the rigidity of the hard-frozen ice-pack beneath them, yet a small rise in temperature could cause it to melt away, leaving them stranded in the sea.

We are entrusting our safety, even on ice that isn't thin, to the organization of the individual molecules. In a crystalline solid it is their orderly arrangement into a lattice that gives the solidity and also the beauty that enraptures: carbon atoms may organize themselves into diamond, or into soot. In a solid the individual atoms are locked in place relative to one another but warmth causes them to jiggle a bit so that each is slightly displaced from its designated location. However, mindful of their neighbours, the positional errors do not accumulate and the whole can retain apparent perfection and solidity. In the liquid phase, the jiggling is so agitated that the atoms break ranks and flow.

In some materials the change is abrupt: a fraction above or below $0\,°C$ can be the difference between life and death on the ice. In others it is not, such as glass, where there is no meaningful way to tell whether it is a solid or a highly viscous liquid. Helium is a gas at room temperature and liquid when cold, but however much you lower the temperature it never freezes. Nonetheless, subject helium to pressure and it will crystallize.

These examples show different phases appearing depending on how these particles organize themselves. Interesting things can happen when the collection reorganizes itself when passing from one phase to another as in the case of water and ice at $0\,°C$.

At any temperature the organized state with the least energy will be most stable and win in any competition to decide the favoured phase. The temperature of a medium is a measure of its energy, especially that due to the kinetic energy of its constituents. The higher the temperature, the greater is the random motion. Below $0\,°C$ the molecules of water tend to lock to one another, their atomic jigsaw forming shapes of crystalline regularity, giving the sixfold fractal patterns familiar in the frost on winter windowpanes. The motion of the molecules at such temperatures is small enough that collisions among them do not have sufficient energy to disrupt the bonds that hold them. However, above $0\,°C$ their energy is higher and the violence of the collisions too great for them to stay linked in crystals of ice. Any piece of ice added to your liquid drink above $0\,°C$ will have its molecules hit so violently by those of the warm liquid that they will break apart from one another and flow as liquid also.

At $0\,°C$ a mixture of liquid and ice will turn to ice as in this phase the molecules have a lower energy than in the liquid state.

As they solidify, the excess energy is released as heat (this is known as the latent heat). The amount is not huge but we could do a gedankenexperiment of imaging what would happen if it was much larger, greater even than the energy to create molecules of ice and 'anti'-ice. If nature had been like this, then as the temperature dropped through $0\,°C$, snowflakes and anti-snowflakes would appear spontaneously, seemingly out of nowhere.

As they do, an interesting enigma would occur. Above $0\,°C$, the ground state of water molecules appears the same in whichever direction we look. We say it is symmetric under rotation. An individual snowflake is not like that. A snowflake has a beautiful shape, a sixfold symmetry such that if you rotate through multiples of 60 degrees you see the same as before, but at any other angle you see a rotated snowflake. A tentacle may point outwards in the 12 o'clock direction, say, forcing the others to be at 2, 4, 6, 8, and 10; or perhaps it is at 1 o'clock, with partners at the odd-numbered positions on the clock face. As billions of snowflakes form, their orientations are random such that overall the new ground state, filled now with snowflakes, appears the same in all directions. However, from point to point the symmetry will be broken; a snowflake is pointed one way here, and another over there.

Another example with important insights for our understanding of the vacuum is the phenomenon of magnetism, which is a result of electrons spinning, each electron acting like a minimagnet. In iron neighbouring electrons prefer to spin in the same direction as one another as this minimizes their energy; to minimize the energy of the whole crowd, all of them must spin in the same direction, which gives an overall north–south magnetic axis to the metal. This is the state of minimum energy, the ground state. However, above about $900\,°C$ the extra energy that the heat

Figure 12. The six-fold symmetry of a snowflake.

provides is more than enough to liberate each of the spinning electrons from the entrapment of its neighbours; in this case these mini-magnets point in random directions and the overall magnetism disappears. So iron can manifest a magnetic phase or a non-magnetic phase, depending on the temperature.

Mythical creatures that lived within such systems would regard the lowest energy state as the background norm. Everything that these creatures perceived about these organized systems would be like what we experience for the vacuum in our universe. Our quantum vacuum is like a medium and never truly empty. It too can be organized in different phases and there are interesting properties and phenomena that can occur as one passes from one phase to another. It is widely suspected that this may have affected the nature of space-time in the early moments of our universe.

So we now have a new perspective on the ancient philosophers' question of whether nature allows a vacuum. The answer is, depending on your point of view, either 'no' (in that the void is actually filled with an infinite sea of particles together with quantum fluctuations) or 'yes; there are many different types of vacuum' (i.e. depending on how the medium that is the quantum vacuum is organized). The received wisdom in physics tends to be in the latter camp. We will learn more about this after seeing how patterns and form can emerge as the quantum vacuum moves from one organized state to another.

Phase Changes and Vacuum

Many physical systems do not show the fundamental symmetries of the forces that build them. Electromagnetic forces don't care about left or right yet biological molecules have mirror images that are inert or even fatal while their originals are food or beneficial.[1]

Balance a perfectly engineered cylindrically shaped pencil on its point. Turn around: it looks the same. This invariance when one rotates is known as a symmetry, in this case rotational symmetry. Balanced on its tip the pencil is metastable as the force of gravity will pull it to ground if it is displaced from the vertical by the slightest amount. The gravitational force is rotationally symmetric, which implies that when the pencil falls to the ground, no particular direction is preferred over another. Do the experiment thousands of times and the collection will show the pencils have fallen to all points of the compass, in accord with the rotational symmetry. However, on any individual experiment you cannot tell in which direction the pencil will fall; having fallen, perhaps to the north, the 'ground state' will have broken the rotational symmetry. Roulette is another example. Play long enough and all the numbers will win with equal likelihood; this guarantees that the house wins as the zero is theirs. But on any individual play it is your inability to predict with certainty where the ball will fall that is the source of the gamble.

In the example of the pencil, the state in which the symmetry is broken is more stable than the symmetric state in which the pencil was precariously balanced on its tip. In general, the laws that govern a system have some symmetry but if there is a more stable state that spoils it, the symmetry is 'spontaneously broken', or 'hidden'. So it was with a snowflake and water or with magnetism of iron.

You may cry foul at this point arguing that this is not really a failure of symmetry, but more a result of one's imprecision in balancing the pencil: 'The pencil dropped because it was not perfectly upright.' This is true, but suppose that it had been balanced

127

on a perfectly engineered point. Even then the atoms in the tip are in random motion, due to the temperature, heat, manifested in their kinetic energy. This randomness means that the direction of toppling is random. You might agree but suggest that we do the experiment at temperatures approaching absolute zero of temperature, $-273°C$, where the kinetic energy tends to vanish. Your gedankenexperiment supposes the tip to be engineered from perfectly spherical molecules, the pivotal one being frozen in place at absolute zero temperature where thermal motion has ceased. The catch is that the quantum laws take over. If motion has vanished, then position is unknown and the point of balance is itself randomized. If the point were precisely known at some instant, motion would be undetermined and the resulting imbalance unpredictable. It seems that here, and in general, the quantum fabric of nature enables high-energy metastability to choose a state of lower energy where the symmetry is spontaneously broken. Thus melting ice, or heating magnetized metal, causes the symmetry to return, but when allowed to cool again, the symmetry is broken with no memory of what happened before.

The rule is that raising the temperature causes structure and complexity to melt away giving a 'simpler' system. Water is bland; ice crystals are beautiful.

The universe today is cold; the various forces and patterns of matter are structures frozen into the fabric of the vacuum. We are far from the extreme heat in the aftermath of the Big Bang, but if we were to heat everything up, the patterns and structures would disappear. Atoms and the patterns of Mendeleev's table have meaning only at temperatures below about $10,000°$; above this temperature atoms are ionized into a plasma of electrons and nuclear particles as in the Sun. At even hotter temperatures, the patterns enshrined in the Standard Model of particles and forces, where the electron is in a family of leptons, with families of quarks and disparate forces, do not survive the heat. Already at energies above 100 GeV, which if ubiquitous would correspond to temperatures exceeding 10^{15} degrees, the electromagnetic force and the weak force that controls beta-radioactivity melt into a symmetric sameness. Theories that describe matter and forces as we see them in the cold imply that all these structures will melt away in the heat. According to theory, the pattern of particles and forces that we are governed by may be randomly frozen accidental remnants of symmetry breaking when the universe 'froze' at a temperature of about 10^{17} degrees. We are like the pencil that landed pointing north, or the roulette wheel where the ball landed in the slot that enabled life to arise. Had the ball landed elsewhere, such that the mass of the electron were greater, or the weak force weaker, then we would have been losers in the lottery and life would not have occurred.

Here I have come full circle back to my starting conundrum. If the spontaneous symmetry breaking had made other parameters and forces, we would not have been here to know it. This has given rise to the radical idea that there may be many vacua, multiplicities of universes, of which ours is the one where by chance the dials were set just right.

An example here is of magnetized metal: heat it, destroying the magnetism, and cool it again. In one part the atomic magnets become frozen together pointing in one direction, while in another part of the metal they lock in another direction. This phenomenon is known as 'magnetic domains'. Could this be a model of the universe? Theorists have built mathematical models of the Big Bang, which have to agree with what we know and exhibit the 'true' symmetry in the early hot epoch. A general feature seems to be that such models imply that when cooling occurs

129

from the initial symmetric state, there is a 'landscape' of possible solutions. When you view the entire landscape, you see on the average the original symmetry: like the orientations of the fallen pencil at all points of the compass, there are all possible masses and forces that are consistent with the original symmetry. What is true hereabouts, and in the billions of light years accessible to us, might be different elsewhere.

Changing Forces in the Vacuum

The effervescence of the vacuum disturbs passing electrons and hence also the forces that one charged particle exerts on another. While the inverse square law of the electrostatic force is natural for electric fields that uniformly spread out through three-dimensional space, precision data show subtle deviations from this. Moving at 1 per cent of the speed of light, the effects of relativity are measurable. The stretching and interweaving of space and time distorts the simple inverse square behaviour giving subtle additional effects that grow more rapidly than the inverse square when two charges approach one another. Most familiar as magnetism, these are the immediate manifestations of relativity. When two charges get even closer, separated by distances smaller than an atom's length, the quantum vacuum further distorts these forces.

As mentioned before, forces are transmitted by particles that carry energy and momentum from one body to another. In the case of the electromagnetic force it is the exchange of photons that does the job. If the photons travel directly from one charged particle to another without disturbance, the inverse square law of force arises; however, when a photon's flight is interrupted by the quantum vacuum, such that it fluctuates into a virtual electron and positron en route, the strength of the force is subtly changed. In effect the negative and positive charges of the virtual electron and positron act like a blanket around the naked charge that spawned the force. Measurements at CERN show that if two charges approach within distances that are some 100 millionth the radius of a hydrogen atom, a thousand times smaller even than the size of its nucleus, the electromagnetic force appears effectively some 10 per cent stronger. Calculations suggest that the strength increases even further at yet smaller distances, though it has not been possible to test this experimentally yet. Modern ideas are that the 'true' strength of the electromagnetic force is perhaps some three times stronger than we perceive in macroscopic measurements. When the electrostatic force causes a comb to attract a piece of paper at a range of a few millimetres, or even when the proton ensnares an electron at atom's length, the force has been enfeebled by the charges of the virtual fields latent within the intervening vacuum. Only at the minutest distances, where all but the most singular fluctuations can intervene, is the true electromagnetic strength to be revealed.

This discovery has given a dramatic change to our view of forces. Within a nucleus there are other forces at work, known as the weak and strong, their names testifying to their strengths as perceived relative to that of the electromagnetic force. The strong force is responsible for holding the positively charged members of the nucleus, the protons, in a tight grip even while their mutual electrical repulsion ('like charges repel') is trying to drive them apart. Within the protons and neutrons themselves, the strong force confines the quarks in permanent imprisonment. One manifestation of the weak force is beta-radioactivity where the nucleus of one atomic element can transmute into another.

As the electromagnetic force is carried by photons, so is the strong force between the quarks carried by gluons, while the weak force is transmitted by electrically charged W bosons or by electrically neutral Z bosons. These different particles are affected by the vacuum in different ways. For example, gluons are blind to electrons, positrons, and photons, but have to force their way through the clouds of quarks and antiquarks, and even other gluons that lurk within the quantum vacuum. The W and Z by contrast feel both charged particles and also the nearly massless electrically neutral particles known as neutrinos and antineutrinos.

Calculations show that while the strength of the electromagnetic force grows as the shielding effects of the vacuum are removed at short distances, the different response of the gluons to the vacuum cause the strength of the 'strong' force to be enfeebled in the analogous circumstances. Experiment has confirmed this. The strong binding forces that grip an atomic nucleus, giving it stability, are thus a result of the vacuum strengthening the gluons' grip at distances of 10^{-15} m. The masses of protons, neutrons, and ultimately of all bulk matter are effectively due to the gluonic vacuum acting over nuclear dimensions. This is surprising, but true. The successful comparisons between data and the calculations, which assume that the quantum vacuum plays an essential role, are too much to be mere accidents. Furthermore, they provide a tantalizing hint that, were it not for the effects of the vacuum, the strengths of all these forces would probably be the same. If true, this implies a profound unity to the forces of nature at source, and that the multitude of disparate phenomena that occur at macroscopic distances, such as our daily experiences, are controlled by the quantum vacuum within which we exist.

To experience the forces and nature at distances so small that the intervention of the vacuum is nugatory requires the study of collisions among particles at exceedingly high energies. Such conditions were commonplace in the early universe where the extreme heat would be manifested by high kinetic energies of the particles. The theory of the forces and the vacuum embodied in the 'Standard Model' of particle physics implies that in the early universe, initially the vacuum state had a symmetric phase where these forces exhibited essentially the same strengths and were in effect unified. As the universe cooled, phase transitions occurred and the symmetric vacuum state was replaced by increasingly asymmetric states. Thus what we now call the strong force separated from the electro-weak, which is the name given to the still unified electromagnetic and weak forces, at a temperature above 10^{28} degrees, which would have occurred around 10^{-34} seconds after the Big Bang.

The separation of the electro-weak into what we now recognize as electromagnetic and weak took place at much lower temperatures, around 10^{15} degrees, which is accessible in experiments at CERN and has been studied in detail. The breaking of this symmetry is rather different from the phase change that had earlier led to the emergence of a separate strong interaction. The 'weak' force appears weak because it is a short-range force, extending over distances smaller than the extent of a proton and hence quite unlike the infinite range of the electromagnetic force. It is its short range that means that its effects at longer range appear feeble even though, close in, its natural strength, essentially the same as that of the electromagnetic force, is revealed. So why is the reach of the weak force so tiny? The answer has to do with the nature of its carriers, the W and Z bosons: whereas the photon is massless, the W and Z are very massive, approaching 100 times the mass of a proton. It is only when the energies of collisions, or temperatures in the universe, are so large as to make the energy locked into the

133

mc^2 of these bosons trifling by comparison, that the unity of the forces is revealed. This brings us to the frontier of current research into the nature of the vacuum, which is concerned with the nature of mass and the Higgs vacuum.

The Higgs Vacuum

The weak force, then, appears feeble because of its limited reach. Compared to the scale of around 10^{-31} m where the forces are unified and the different effects of the quantum vacuum are nugatory, the 10^{-18} m range of the weak force is so large as to be effectively infinite. In energy terms, whereas the photon has no mass, the W or Z carriers of the weak forces have masses approaching 100 GeV. Even though this is still small compared to the effective energy scale of 10^{16} GeV at which the unification range is resolved, it nonetheless begs the question of how these W and Z bosons can acquire mass while the photons and gluons, to which they are supposedly related, have none. The answer is believed to be due to a property of the vacuum, which is currently at the frontier of research in high-energy particle physics.

The theory, which is due to Peter Higgs, builds on ideas about superconductivity suggested by Philip Anderson, according to which the photon acts as if it has become massive. Superconductivity, as its name suggests, is the property of some solids to lose all resistance to the flow of electric current when the temperature falls low enough. This change from being a relative insulator to being a superconductor is an example of a phase change. But there is more to superconductivity than just the sudden freedom of electrons to flow; there is also what is known as the Meissner effect, which relates to how magnetic fields behave in and around a superconductor. A magnetic field may permeate a warm solid, but at low temperatures, where the material becomes superconducting, the magnetic field is abruptly expelled from all but a thin skin at the surface. Inside the solid, the magnetic field reaches only over a limited distance, x; if we recall how the limited range of the weak force correlates with the mass of the carrier, the W, so within the superconductor the short range of the magnetic field is as if its carrier particle, the photon, has gained a mass of the order of $\hbar/xc$.

The theory of this phenomenon is profound and entire books could be written about it[2]; it is not my intention to do so here. By analogy, applied to the weak force, we want the W field of mass M to be able to penetrate the physical vacuum by only a distance $x = \hbar/Mc$. The jargon is that the physical vacuum as perceived by the weak force acts like a superconductor.

The phenomenon of superconductivity depends on the existence of matter fields with special properties. In a real superconductor the expulsion of the magnetic field arises as a result of electrons within the material acting cooperatively giving what are known as 'screening currents'. In the case of the weak force the analogy requires that there must be some matter field present *in the vacuum*. This is profoundly different from what we have met so far. Hitherto we have contemplated the quantum vacuum filled with virtual fields, fluctuations about zero that can only be materialized if further energy is supplied. But now, with the 'Higgs field', we are contemplating something that has a genuine presence in the vacuum: the 'empty' space with no Higgs field would have *more* energy than when the Higgs field is present. Put another way: add a Higgs field to the void and the overall energy is reduced.

135

This surprising result also has analogues in solids, such as magnets, as we saw on p. 129. Above some temperature, known as the 'curie temperature' T_c, the metal has the least energy when it is not magnetic; however when cooled below T_c the metal becomes a magnet. Thus at low enough temperatures, 'adding' magnetism lowers the energy of the ground state, or 'vacuum'.

The favoured theory in particle physics is that the Higgs field pervades the vacuum and gives mass to the fundamental particles, not just to W and Z bosons but to electrons, quarks, and other particles too. If this is true then in the absence of the Higgs field particles could never be stationary but would all travel at the speed of light. However, space is filled with the Higgs field. As you read this page you are looking through the Higgs field: photons do not interact with it and move at the speed of light.

The Higgs field is indeed bizarre. Particles such as electrons travelling through space at speeds below that of light are doing so because they have mass, which they have gained as a result of interacting with the omnipresent Higgs field. Yet they continue to travel without resistance: Newton's laws work, the particles continuing to move at constant velocity as no external force appears to act on them. A partial answer to this conundrum comes if we realize that a particle's energy determines its velocity; as the Higgs field is the vacuum state of lowest energy, no energy can be transferred by the particle to or from the Higgs field, and so the particle maintains its speed. It is not possible to determine an absolute value of the velocity relative to the Higgs field.*

As superconductivity and magnetism are the lowest energy states only at low enough temperatures, so is the Higgs-permeated vacuum the lowest energy state only at sufficiently 'low'

---

* In the technical jargon: 'The Higgs vacuum is a relativistic vacuum.'

temperatures, where 'low' means 10^{17} degrees! At temperatures above 10^{17} degrees, theory suggests that the ground state of the universe does not include the Higgs field. For the first trillionth of a second after the Big Bang the universe was hotter than this and it is only since that time that the Higgs field has filled the void, giving masses to the fundamental particles.

As ripples in electromagnetic fields produce quantum bundles, photons, so should the Higgs field manifest itself in Higgs bosons. In a chicken and egg manner, the Higgs boson itself feels the all-pervading Higgs field and so has mass. Higgs's theory implies that the eponymous boson has a huge mass, up to a thousand times that of a hydrogen atom. Quantum uncertainty implies that virtual Higgs bosons are fluctuating in and out of the vacuum and precision measurements of how the vacuum affects the motion of particles such as electrons, and the properties of the force carriers W and Z, suggest that these are affected by the virtual Higgs bosons too. When all the data are compared, it appears that the Higgs boson may be lighter than previously thought, perhaps only 150 times the mass of a hydrogen atom. At CERN a 27-km ring of magnets can steer beams of speeding protons, which when smashed into one another head-on can produce the conditions suitable for the Higgs bosons to be produced. This accelerator, known as the Large Hadron Collider or LHC, took ten years to build and was finally completed in 2007. The experiments may take months to perform and years to analyse and refine. If the void really is filled with a Higgs field, we should know this very soon.

9

THE NEW VOID

The Start of the Universe

One hundred and thirty-seven pages ago we started with the question 'where did everything come from?' Having surveyed over 2,000 years of ideas, we have arrived at the modern answer: 'Everything came from nothing.' Modern physics suggests that it is possible that the universe could have emerged out of the vacuum. 'There could hardly be a more remarkable interconnection than this between "nothing" and "something"',[1] or more colloquially 'The universe may be the ultimate free lunch.'[2] The idea is that our universe could be a gigantic quantum fluctuation with total 'virtual' energy so near to zero that its lifetime can be huge. This can occur because there are both positive *and* negative energies in the universe due to the all-pervading attraction of gravity. To illustrate this it is perhaps easiest first to recall how the electric forces within an atom discouraged invaders in Chapter 2.

138

The atomic nucleus being positively charged is surrounded by an electric field that repels other positive charges such as alpha particles. Imagine such an alpha particle, very far from the nucleus and speeding towards it. The total energy is then the kinetic energy of the alpha particle.* The nearer that the alpha particle approaches the nucleus, the more it will feel the electrical repulsion and be slowed. If it is on a direct collision course it will eventually momentarily come to rest before being ejected back along the original path. At the instant where it has stopped, its kinetic energy is zero. The total energy must be conserved; we say that kinetic energy has been exchanged for potential energy.

When the strength of a force varies as the inverse square of distance, as here, the magnitude of the potential energy varies inversely with distance. Thus when this distance is large, as at the start of the alpha particle's journey, the potential energy is near to zero. The nearer that the alpha particle approaches the nucleus, so the greater is its potential energy. This grows as the kinetic energy falls until at the moment of closest approach, where the particle is momentarily at rest, the potential energy is maximal and with a magnitude that equals the amount of kinetic energy that it had at the start.

In this example all the energies are positive; a positive kinetic energy at the start is transformed into positive potential energy as it approaches the nucleus. Now suppose that instead of two positive charges, one was negative, such as if a remote electron is attracted towards a positively charged nucleus. If the remote electron were initially at rest its kinetic energy would be zero and, being far from the nucleus, so would its potential

---

* For simplicity I am ignoring relativity and the mc^2; the essential conclusions are unaffected.

energy. The sum of kinetic and potential energies in this case is effectively zero. But as we have an *attractive* force here, the electron is pulled towards the nucleus, gaining speed and hence kinetic energy. As the total must remain at zero, the increasingly positive kinetic energy implies that the potential energy must be negative, and ever more so as the electron approaches the nucleus. So for an attractive force, the potential energy can be negative.

This will be true for gravity where masses attract one another. The potential energies of the Earth or the planets trapped in the gravitational field of the Sun are uniformly negative. Indeed, the *sums* of their kinetic and potential energies are less than zero, which is why they are bound in the solar system, trapped in the Sun's gravitational field. Likewise, you and I are trapped in the earth's gravitational field. Propel an object upwards with kinetic energy and it will fall back to ground unless you give it a starting speed greater than about 12 km per second, known as the 'escape velocity'. Only above such speeds does the sum of the kinetic energy and the potential energy become positive whereby something can escape from the Earth's gravity; however you will still be trapped within the solar system with a total energy that is negative.

The attractive force of gravity pervades the cosmos with consequent negative potential energy for everything trapped within it. It is possible that even when all of the mc^2 of its matter is included, the total energy of the universe is near to nothing. Thus, according to quantum theory, the universe could be a huge vacuum fluctuation where the total energy is so near to zero that it can exist for a very long time before the vacuum accountant demands that the books are balanced. If the total energy is zero, it could last for ever.

If this is so, then who is to say that ours is the one and only universe? Why not allow the possibility that there are other bubbles of effervescing multiverses? Many theorists seriously contemplate such a possibility though many argue about whether such ideas are within science in the sense of being accessible to experimental test.

As the universe expands, space extends but objects held together by electromagnetic forces, such as planets and stars, do not change size; the space between them grows. Electromagnetic radiation has nothing to hold it and so its wavelength extends as the universe grows. From quantum theory we know that the wavelength correlates inversely with energy, so the cosmic background radiation, which is today a mere 3° above absolute zero, would in the past have been much hotter. Similar conclusions follow for matter. As the universe expands, matter trapped in the all-pervading gravitational field will have its potential energy grow at the expense of its kinetic energy. This universal slowing is perceived as a fall in temperature. So from the observed rate of expansion of the universe and knowledge of its background temperature now, we can calculate back and estimate its temperature at various epochs in the past. It gets ever hotter as we approach nearer to the singular event that we call the Big Bang.

Collisions among particles would have been much more violent then, so much so that at temperatures greater than 4,000° atoms would be unable to survive; they would have been ionized as they are within the hot Sun today. At temperatures above a billion degrees even atomic nuclei are disrupted; a plasma of particles and radiation is all that would have existed in those first moments. Prior to this the energy would have been enough for particles of matter and antimatter to have emerged. All the evidence implies that our material universe came from a vacuum of hot radiation.

Experiments at particle accelerators, such as at CERN, show how matter particles and forces behave at high energy and, by implication, at extreme temperatures. This enables us to calculate how the universe behaves all the way back to temperatures of 10^{27} degrees, which corresponds to times within 10^{-33} seconds of the Big Bang. As we saw earlier, at various temperatures the vacuum undergoes phase transitions, some of which have been established, others being theorized. As it cooled beneath 10^{15} degrees after about 10^{-10} seconds, the electromagnetic and weak forces took on their separate identities; this has been established by recreating these energies experimentally. Theory predicts that at slightly higher energies, corresponding to time scales of around 10^{-12} seconds, the cooling vacuum had made a phase transition where the Higgs field froze in and particles gained masses.

So we have a picture of the universe erupting as a quantum fluctuation in the vacuum, which somehow is exceedingly hot and expanded rapidly. This picture would have led to vast amounts of matter and antimatter being produced symmetrically, yet there is no evidence for antimatter surviving in bulk today. It is generally believed that there has to be some asymmetry between protons and antiprotons. The origin of this is still being sought but it may be another example of spontaneous symmetry breaking as the universe went through a phase change.

Inflation

Problems remain with this scenario including the question of where all the thermal energy came from. Furthermore, from experience with phase changes in condensed-matter physics we know that they are never completely smooth. For example, when hot metal cools to make a magnet, the magnetism varies from one region to another, forming 'domains' of distinct magnetism. There are defects, non-uniformities throughout the metal. The same ought to have happened throughout the universe when it underwent phase transitions, giving phenomena such as walls of energy, cosmic strings, call them what you will. In any event, there has been no clear sighting of any such bizarre entities. Also, theory suggests that such a sequence of events would have led to the universe's evolution being so fast that its lifetime would have been little more than some tens of thousands rather than the present tens of billions of years. A possible solution to these paradoxes came with the idea of Alan Guth and Paul Steinhardt[3] who proposed that the universe is a domain in some bigger omniverse. In this theory, known as inflation, our universe is the result of an enormous swelling of a single one of these microscopic 'domains'. At first sight this seems impossible as it requires matter spontaneously to fly apart, which ought not to happen when there is a universal gravitational attraction at work. However, in general relativity not just the mass energy and momentum but also the pressure contribute and if the pressure were negative, and dominant over the matter and thermal energies, the result could be a rapid expansion, a sort of 'anti-gravity' effect.

What Alan Guth had noticed was that if the true vacuum contains a Higgs field, there is the possibility that a region of the universe could have been in the unstable or 'false' vacuum. (The false vacuum is akin to the state of the pencil balanced on its point and the true vacuum is the pencil fallen to the table.) Recall that adding the Higgs field to the false vacuum will *lower* the energy. In the false vacuum the total energy is proportional to the volume, and it requires work to increase that volume. Due to the lower energy state in the Higgs vacuum, the natural tendency will be for such a volume to contract, and with respect to the true Higgs vacuum the false vacuum state will be one in which the pressure is effectively negative. So if a fluctuation occurs in a region of false vacuum, the gravitational effect of the negative pressure can overwhelm that of the matter, leading to an expansion. As the universe makes the transition from the false to the Higgs vacuum in this picture, it is possible that a huge inflation can occur in a remarkably short time.

There are examples in condensed-matter physics of super-cooled systems. This is where the system remains in the 'wrong' phase such as when water stays liquid below the nominal freezing point. A similar thing could have happened with the vacuum of the universe. A fluctuation occurs in the false vacuum and continues; only later does the universe make the transition to the true vacuum. It has been calculated that in such circumstances a region of space could double in size every 10^{-34} seconds![4]

After the period of inflation, the transition to the true vacuum releases energy, like the emission of latent heat when water freezes. This energy produced the particles of matter that would eventually form the galaxies of stars, and us. The gravitational attractions give a negative potential energy that cancels this, leaving the total energy near to zero.

It is surprising what an effect this inflation can have. Our present observable universe is some 10^{26} m across. Extrapolating backwards, at temperatures of 10^{28} degrees, which is when the inflation ended, our future universe would have been some centimetres in size. The inflation era would have expanded it by as much as 10^{50} which would imply a starting bubble fluctuation of only 10^{-52} m, which is well into the size of fluctuations that one would expect in quantum gravity.

During inflation there was a runaway expansion that took place ever faster. This was so rapid that some objects that had been near enough to exchange information, such as radiation, would have been thrust into disconnected parts of the universe whereby they are too far apart to exchange information now. For example, there are galaxies some 10 billion light years away from us in opposite regions of the sky, which means they are currently separated by more than 14 billion light years, greater than the distance that light could travel in the lifetime of the universe. Yet these galaxies exhibit the same laws of physics, the spectra of their elements being like fax messages from afar that reveal the elements and their properties to be the same throughout the observable universe. The background radiation across the universe has temperature and intensity the same everywhere to better than one part in 10,000. That this uniformity is total chance is stretching credulity. All of our observable universe must have been causally connected at some time in the past; in the absence of inflation this would be a paradox.

There is a lot of mathematical work going on concerning how fields can behave in quantum theory. One conclusion of this is that inflation is almost unavoidable, which is good news for explaining the universe but makes it hard to pin down the precise mechanism. All we can at present do is to work backwards from what we now see in the universe and compute what inflation must have been like, then see if we can test the consequences. Small fluctuations in space-time structure in the era of quantum gravity act as attractors for collecting matter, which eventually grow to seed the galaxies of stars. If we make a computer simulation of the behaviour of the universe, taking into account its present structure and gravity, and play it backwards in time we find that vacuum fluctuations must have been about 1 in 10,000

Figure 13. Variations in the cosmic microwave background radiation as seen by the COBE and WMAP satellites.

in intensity. The exciting implication is that these would have been present in the background radiation before the galaxies formed. In the last few years this has been dramatically confirmed by precision measurements of the radiation by satellites: the COBE (Cosmic Background Explorer) and WMAP (Wilkinson Microwave Anisotropy Probe). These show variations at the level of a few parts in 10,000 in the temperature. In particular they measure these fluctuations at various resolutions, small angles, or larger spreads, and find a fractal behaviour—the finer the resolution, so the more detail keeps showing up in a repetitive fashion. These phenomena appear to be in accord with what is expected if they are the aftermath of inflation. Small wonder that the Nobel Prize for physics in 2006 was awarded to the leaders of this research.

So our best data are consistent with the theory that our large-scale universe erupted through inflation. If this is the case, we have a possible answer to the question of where we came from. This is all consistent with our picture of the universe based on observation and experimental science. While it provides answers to my original question, the price is that it raises yet more questions that are potentially even more profound. Inflation followed after the era when gravity dominated. We have alluded to the bizarre properties that space-time would have had with fluctuations in the metric and have even seen that the background radiation shows what appear to be fossil relics of such fluctuations. There is no reason to believe that our inflationary universe is, was, a one-off event. There could be many other such universes that have erupted in similar fashion to this but which are beyond our awareness. When confronted by the astonishing range of coincidences in the nature of the forces, the masses of the fundamental particles, even in there being three dimensions of space, but for which the conditions for life would have almost certainly not have arisen,[5] one is forced to wonder why our universe has turned out so conveniently for us. One line of conjecture among scientists is that there are multiple universes, potentially an infinite number, with their own parameters and dimensions; one of these happens to be just right for life, and that is where we have evolved. So welcome to the multiverse, though I am sceptical whether such conjectures can be tested within the realms of science.

Higher Dimensions

In the philosophy of a quantum universe, what we call space and time emerged out of a quantum bubble. There is nothing in known science that runs counter to this philosophy and it accommodates much that occurs in our observable universe, but on the other hand there is no agreed mathematical description of it that inexorably leads to the universe as we perceive it and there are no definitive experimental tests. So for now it is a statement of faith but as experimental techniques improve there may be more that will come into the domain of experimental science. Within such caveats, let us continue.

One popular theory is that there are more dimensions in the universe than we are currently aware of. In the jargon some of these 'curled up' into scales of such small size that they are beyond experience, whereas others expanded from the Big Bang to form the macroscopic large-size familiar ones in our universe of four-dimensional space-time. Which begs the question of what actually dimensions are, whether they exist in the absence of 'stuff', what is special about three spatial dimensions, and if there are further dimensions, how can we reveal them scientifically?

If you were only aware of one dimension, the Greenwich meridian running north–south for example, then someone who headed off eastwards would disappear from your linear world. Were we aware only of a surface, aircraft would literally take off from view. If there were a fourth dimension, and some super-creatures were able to move about in it, then they could appear in our world and disappear again from it like phantoms as their four-dimensional path crossed through our three-dimensional home. This almost reaches the limits of our real world. Moving into a fourth dimension and disappearing may sound like science fiction but we can imagine it: for example, time, and the much beloved science fiction genre of time travel.

In what sense is time a dimension? It certainly has extent in that history records dates as if they are points along some direction. Were someone in our three-dimensional space able to make a time machine such that they could move off to a different point in the time direction, forwards or backwards at will, the time dimension would appear no different from one of space. If we too had such a machine, we could move through this fourth dimen-sion with them; without such a machine we are stuck in the ever present now and, as the time traveller moved off to 'yesterday', they would disappear from our view. Conversely, had we been in the right place in the three-dimensional space yesterday, we might have witnessed the sudden appearance of people out of nowhere.

So time certainly has the character of a dimension, but one that is qualitatively different from those of 3D space. We exist at a point in time, 'now'; and now is another now, at which point we can recall what was occurring at the previous now, while nows yet to come, have yet to come. It is a dimension that has a limit; we can look back in time by looking out into space, for light takes time to travel from distant stars to here.

149

As we look at the Moon we see it as it was a heartbeat ago; the Sun as it was eight minutes ago; the light from stars in the night sky has travelled across the intervening space for thousands of years or more, so we are seeing them as they were back then. If some creature on a planet encircling one of those distant stars is looking into their own night sky as you read this, they will see our Sun as it was thousands of years ago, perhaps even before humans walked the Earth. The concept of 'now' becomes less clear.

We could look deep into space and back in time, towards the time we call the Big Bang. If our universe of space and time came into being then, the time dimension has an extent of no more than 14 billion years so far. We can travel on into that future and can look back into its past; thus time has the character of a dimension through which we can move, but it differs from those of 3D space. If we do not like where we have come to in space, we can return whence we came; in time we cannot.

Ever since Edwin Hubble observed that the galaxies are rushing away from one another, we have had the picture of an expanding universe, the 14 billion-year-old remnant of the event that we call the Big Bang. What caused the Big Bang, from where and whence it came, is the modern version of all creation myths. An advantage for us in attempting to answer this is that we know of the quantum, which gives new insights, not just into the idea that the vacuum is a medium but also into the uncertain nature of space and time. The best that we can do is to imagine that epoch from the perspectives of our own experiences. These tell us that when matter is packed tightly together it feels the force of gravity, which is subject to general relativity and the laws of quantum mechanics.

As we have seen earlier, while general relativity describes the cosmos at large, and quantum mechanics makes precision statements about phenomena on subatomic length scales, a mathematically consistent and experimentally tested theory that combines these two great pillars of twentieth century science has yet to be achieved. In order to understand what the universe was like during the very first moments of the Big Bang we require such a quantum theory of gravity. In general relativity the curvature of space-time is related to the concentration of energy. The uncertainty or smearing of energy in the quantum universe will lead to a consequent uncertainty in the curvature of space-time, leading to fluctuations in the distances, or more specifically, the metric (p. 87). The entire geometry will be uncertain; the notion of dimensions and even their number pale beyond our experience.

The most promising current theories that address these problems appear to work best if the universe has many dimensions, perhaps ten, and are known as string theories.[6] Whether this simplification of physics at high energies in many dimensions is a mathematical trick enabling sensible calculations to be made, or whether it is hinting at something profound about the fabric of the universe, is currently unresolved. In any event, in order to relate such a higher-dimensional universe to the one that we perceive, we must assert that all but three of the space dimensions are imperceptibly small. While all these dimensions were symmetrically important in the era of quantum gravity, only space and time as we know them inflated to accommodate the macroscopic universe that we are aware of today.

Whether there are dimensions beyond those we call space and time may be answered scientifically soon. In addition to up–down, front–back, and sideways, there could be further directions 'within'. Until very recently it was thought that these higher dimensions were lost forever in the quantum foam, but novel ideas that attempt to explain why the force of gravity is so much weaker at the atomic scale than the other forces have suggested that gravity may be leaking into a higher dimension that could even exist at scales accessible to experiments at CERN's new LHC (Large Hadron Collider). Earlier we gave the example of a plane taking off in the third dimension apparently disappearing from the view of a two-dimensional flatlander; analogously, particles appearing from the fifth dimension, or disappearing into it, could be a signal at the LHC that space-time is indeed, like Emmenthal cheese, permeated with little bubbles which are at the edge of our present abilities to measure.

In Search of The Void

The idea of higher dimensions may be reality, or it may be science fiction, but it is in any event a helpful aid to the mental gymnastics required when trying to solve the paradoxes associated with where the universe was the day before creation.

Our problem is tied to our perception that time is a simple linear dimension. If the history of the universe is listed on a vertical line, 'now' is some point on it, the future lies above and the past lower down, the Big Bang at the bottom. But there the line stops; below this is nothing. In this linear timescape, there was no time before time. This is where the concept of the Void has flourished; where poetry fills the vacuum of our incomprehension at the limits of imagination. For the author of Genesis, in the beginning there was 'darkness on the face of the deep'; in the Rigveda the unknown is even more profound: 'darkness was hidden by darkness'.

We have met the idea of flatlanders who are unaware of the dimension out of the page. Perhaps we are like them, unaware of dimensions beyond the familiar.

We have already seen Einstein with his four-dimensional picture of space-time, with curvature related to gravity. Hawking and Hartle have gone further and imagine the universe as a four-dimensional surface of a five-dimensional sphere. I cannot visualize this, nor to be fair do its authors other than mathematically. However, we can visualize a simpler version, playing once again the role of creatures who are aware only of limited spatial dimensions whose universe is perceived to be expanding in time. This will suggest that our universe only appears to be expanding as a result of our limited cognition. In the Hawking–Hartle model, there is no expansion, no beginning: the universe simply exists.

Instead of our three space dimensions plus time, imagine a universe with just one dimension plus time, heading from a single point (its 'Big Bang') to an end point (its 'Big Crunch'). Hawking and Hartle have suggested that time might not be a simple linear flow but has another dimension, which they call 'imaginary time'. Suppose that we represent the universe with one spatial dimension, and time plus imaginary time as the surface of a sphere. We can identify points on this ball by their latitude and longitude, much as we do on the surface of the Earth. In Hawking and Hartle's picture, the lines of latitude are the coordinates of time, and the longitude is what they call 'imaginary time'. The Big Bang is then at the north pole and the Crunch at the south pole. Each line of latitude corresponds to a particular time, for example 40 degrees north might represent 'today'.

Now look at the region near the north pole. As we approach time zero, the grid for imaginary time becomes condensed, much as approaching the north pole makes all lines of longitude

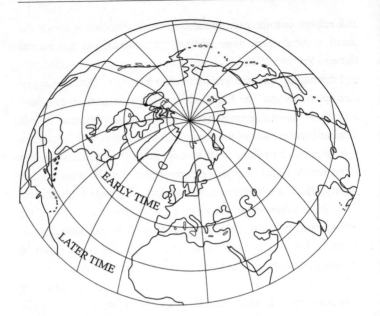

Figure 14. The history of the universe in space and imaginary time.

converge. There is nothing singular about the pole; the fact that all lines of longitude converge there is just an 'accident' of how we chose to draw the grid. On our globe you can imagine travelling around the Arctic and, apart from being cold, it is no different from travelling anywhere else on the surface. We could have layered the globe with lines radiating out from London and homing in on the Antipodes if we had wished.

Possibly Hawking and Hartle's imaginary time is just that— imaginary. Or maybe this is a mathematically consistent theory that is just beyond imagination. It is the modern example of the problem that has plagued three millennia of thinkers: our minds have developed a view of the world based on our macroscopic sense of time and three space dimensions. We describe matter and energy within this mental construct. Paradoxes about the 'start' of the universe arise when we are restricted to this mental picture. Yet 14 billion years ago space and time were so warped and fluctuating that the 'reality' would have been far beyond our conceptual ability. The Big Bang created space and time. Before it ('before' of course only having meaning in the sense of our familiar mental matrix) there was no yesterday.

It is possible to imagine that what we call the Big Bang was when the compact universe emerged from the era of quantum gravity, which is when time took over from imaginary time. Questions about where everything came from, how it all 'began', are sidestepped; the universe in this picture has no beginning, no end: it just is. Do you feel that this is the answer to the question of the ages, that the paradox of creation has been resolved? I am not convinced; imaginary time is, for me at least, unimaginable. We may have given a name to the big question but that is not the same as understanding the answer. Why the universe is, and in what, remain enigmas.

If multiple universes have erupted as quantum fluctuations, such that our bubble happens to have won the lottery where the laws, dimensions, and forces are just right for us to have evolved, this still begs the question of who, what, where were encoded the quantum rules that enable all this. Was Anaxagoras right: the universe emerged as order out of chaos, the ur-matter is the quantum void? Or perhaps Hawking and Hartle's conception of a universe that has no beginning or end, and simply exists, is the answer, such that Thales, who insisted that something cannot come from nothing, is right. The paradox of creation is thus an as yet unresolved mystery about the nature of space and time.

In the 3,000 years since the philosophers of ancient Greece first contemplated the mystery of creation, the emergence of something from nothing, the scientific method has revealed truths that they could not have imagined. The quantum void, infinitely deep and filled with particles, which can take on different forms, and the possibility of quantum fluctuation lay outside their philosophy. They were unaware that positive energy within matter can be counterbalanced by the negative sink of the all-pervading gravitational field such that the total energy of the universe is potentially nothing; when combined with quantum uncertainty, this allows the possibility that everything is indeed some quantum fluctuation living on borrowed time. Everything may thus be a quantum fluctuation out of nothing.

But if this is so, I am still confronted with the enigma of what encoded the quantum possibility into the Void. In Genesis some God said, 'let there be light,' but for the Rigveda, gods are creations of human imagination, invoked to explain what lay beyond understanding: 'the Gods came afterwards...who then knows whence all has arisen?' As science discovers answers, it exposes deeper questions whose answers are for the future. In the meantime, I leave you with a poetic interpretation from the Rigveda:

> The non-existent was not; the existent was not
> Darkness was hidden by darkness
> That which became was enveloped by The Void.

Notes

Chapter 1

1. Comment by Richard Dawkins in television interview, 2005.
2. There are many translations of the Rigveda as a search on Google will reveal. These particular lines are from the translation by Wendy Doniger O'Flaherty in *The Rig Veda Anthology* (Penguin, 1982).
3. B. Pascal, posthumous notes quoted in H. Genz, *Nothingness* (Perseus, 1999).

Chapter 3

1. See Genz, *Nothingness*, 85.
2. Unattributed quote, ibid. 110.
3. See Close, *Lucifer's Legacy*, 21–7.

Chapter 5

1. As reported in Hey and Walters, *Einstein's Mirror*, 50.

Chapter 7

1. As quoted in *Science*, 256 (1992), 1519.
2. Laughlin, *A Different Universe*, 47.
3. For further pictures of pair creation and other examples of such quantum effects, see F. E. Close, M. Marten, and C. Sutton, *The Particle Odyssey* (Oxford University Press, 2002) and for descriptions of how such images are obtained see also Frank Close, *Particle Physics: A Very Short Introduction* (Oxford University Press, 2004).

Chapter 8

1. The many bizarre asymmetries between left and right are described in Close, *Lucifer's Legacy*, and C. McManus, *Right Hand, Left Hand* (Weidenfeld and Nicolson, 2002).

2. See I. Aitchison, Nothing's Plenty', *Contemporary Physics*, 26 (1985), 333.

Chapter 9

1. Aitchison, 'Nothing's Plenty', 385.

2. A. Guth, as quoted in S. Hawking, *Brief History of Time* (Bantam, 1988); for a definitive description see V. J. Stenger, '*The Universe: The Ultimate Free Lunch*', *European Journal of Physics*, 11 (1990), 236.

3. A popular description of inflation theory is in *Scientific American*, 250 (1984), 116. See also J. D. Barrow, *The Book of Nothing* (Vintage, 2000), 255 and Aitchison, 'Nothing's Plenty'.

4. Aitchison, 'Nothing's Plenty', 388.

5. There are many books and articles on the Anthropic Principle as a search on Google will reveal. A recent one, which touches on the ideas of multiple universes, is by P. Davies, *The Goldilocks Enigma* (Allen Lane, 2006).

6. Such theories are mathematically profound and exciting. See Brian Greene, *The Elegant Universe* (Jonathan Cape, 1999). However, it is far from clear whether they are primarily adventures in mathematics or the long-sought theory of everything. For a critical evaluation see also Peter Woit, *Not Even Wrong* (Jonathan Cape, 2006).

Bibliography

There is much about nothing that I have been unable to include here, and much more that has already been written. I have referred to some of these books and articles in the text, and collect them here together with some suggestions for further reading. This is by no means exhaustive. If you are seriously interested in nothing, the books by Barrow and Genz in particular contain an extensive list of references and original sources.

The Book of Nothing by John D. Barrow (Vintage, 2000) and *Nothingness* by Henning Genz (Perseus, 1999) go further and more deeply in some cases into the story of the vacuum and other manifestations of 'nothing'. Barrow discusses also the mathematical story of zero and aspects of cosmology, in particular of multiple universes, in detail. Genz has a particularly good description of the Higgs mechanism and of spontaneous symmetry breaking in condensed matter systems.

A Different Universe by Robert Laughlin (Basic Books, 2005) describes the emergent nature of the laws of macroscopic phenomena and of the nature of the vacuum.

Lucifer's Legacy by Frank Close (Oxford University Press, 2000) describes spontaneous symmetry breaking and many examples of symmetries in Nature. *Particle Physics: A Very Short Introduction* (Oxford University Press, 2004) and *The New Cosmic Onion* (Taylor and Francis, 2007), both by Frank Close, give the ideas of particle physics that form some of the background to the later chapters of the present book.

The Goldilocks Enigma by Paul Davies (Allen Lane, 2006) describes the ideas of multiple universes and how our particular universe is so finely tuned for life.

The Particle Odysseys by F. E. Close, M. Marten, and C. Sutton (Oxford University Press, 2002) is a highly illustrated history of modern physics.

Einstein's Mirror by A. Hey and P. Walters (Cambridge University Press, 1997) gives a popular introduction to relativity and *The New Quantum Universe* (Cambridge University Press, 2003) does the same for quantum theory.

'Nothing's Plenty: The Vacuum in Modern Quantum Field Theory' by I. J. R. Aitchison, in *Contemporary Physics*, 26 (1985), 333 gives a more advanced discussion of modern ideas about the quantum vacuum.

Index